Bayesian Networks and Bayesian

A Practical Introduction for Researchers

Stefan Conrady

Lionel Jouffe

Bayesian Networks and BayesiaLab—A Practical Introduction for Researchers
Copyright © 2015 by Stefan Conrady and Lionel Jouffe

All rights reserved. No part of this publication may be reproduced, distributed, or transmitted in any form or by any means, including photocopying, recording, or other electronic or mechanical methods, without the prior written permission of the publisher, except in the case of brief quotations embodied in critical reviews and certain other noncommercial uses permitted by copyright law. For permission requests, write to the publisher:

    Bayesia USA
    312 Hamlet's End Way
    Franklin, TN 37067
    www.bayesia.us
    info@bayesia.us
    +1 (888) 386-8383

Ordering Information:
Special discounts are available on quantity purchases by corporations, associations, and others. For details, contact the publisher at the address above.

ISBN: 978-0-9965333-0-0

Contents

Preface	ix
Structure of the Book	x
Notation	xi
1. Introduction	13
All Roads Lead to Bayesian Networks	13
A Map of Analytic Modeling	15
2. Bayesian Network Theory	21
A Non-Causal Bayesian Network Example	23
A Causal Network Example	23
A Dynamic Bayesian Network Example	24
Representation of the Joint Probability Distribution	25
Evidential Reasoning	27
Causal Reasoning	27
Learning Bayesian Network Parameters	29
Learning Bayesian Network Structure	29
Causal Discovery	29
3. BayesiaLab	33
BayesiaLab's Methods, Features, and Functions	34
Knowledge Modeling	35
Discrete, Nonlinear and Nonparametric Modeling	36
Missing Values Processing	37

Parameter Estimation	37
Bayesian Updating	38
Machine Learning	38
Inference: Diagnosis, Prediction, and Simulation	41
Model Utilization	45
Knowledge Communication	46

4. Knowledge Modeling & Reasoning — 49

Background & Motivation	49
Example: Where is My Bag?	51
Knowledge Modeling for Problem #1	52
Evidential Reasoning for Problem #1	60
Knowledge Modeling for Problem #2	67
Evidential Reasoning for Problem #2	73

5. Bayesian Networks and Data — 79

Example: House Prices in Ames, Iowa	79
Data Import Wizard	80
Discretization	84
Graph Panel	90
Information-Theoretic Concepts	95
Parameter Estimation	99
Naive Bayes Network	105

6. Supervised Learning — 113

Example: Tumor Classification	113
Data Import Wizard	115
Discretization Intervals	119
Supervised Learning	123
Model 1: Markov Blanket	123
Model 1: Performance Analysis	126
K-Folds Cross-Validation	129
Model 2: Augmented Markov Blanket	133
Cross-Validation	136

Structural Coefficient	138
Model Inference	144
Interactive Inference	146
Adaptive Questionnaire	147
WebSimulator	151
Target Interpretation Tree	156
Mapping	160

7. Unsupervised Learning — 165

Example: Stock Market	165
Dataset	166
Data Import	168
Data Discretization	170
Unsupervised Learning	176
Network Analysis	179
Inference	186
Inference with Hard Evidence	187
Inference with Probabilistic and Numerical Evidence	188
Conflicting Evidence	194

8. Probabilistic Structural Equation Models — 201

Example: Consumer Survey	201
Dataset	202
Workflow Overview	202
Data Import	203
Step 1: Unsupervised Learning	207
Step 2: Variable Clustering	216
Step 3: Multiple Clustering	227
Step 4: Completing the Probabilistic Structural Equation Model	248
Key Drivers Analysis	253
Multi-Quadrant Analysis	266
Product Optimization	273

9. Missing Values Processing — 289

- Types of Missingness — 290
- Missing Completely at Random — 291
- Missing at Random — 293
- Missing Not at Random — 295
- Filtered Values — 296
- Missing Values Processing in BayesiaLab — 298
- Infer: Dynamic Imputation — 311

10. Causal Identification & Estimation — 325

- Motivation: Causality for Policy Assessment and Impact Analysis — 326
- Sources of Causal Information — 327
- Causal Inference by Experiment — 327
- Causal Inference from Observational Data and Theory — 327
- Identification and Estimation Process — 328
- Causal Identification — 328
- Computing the Effect Size — 328
- Theoretical Background — 328
- Potential Outcomes Framework — 329
- Causal Identification — 330
- Ignorability — 330
- Example: Simpson's Paradox — 332
- Methods for Identification and Estimation — 334
- Workflow #1: Identification and Estimation with a DAG — 334
- Indirect Connection — 336
- Common Parent — 337
- Common Child (Collider) — 337
- Creating a CDAG Representing Simpson's Paradox — 338
- Graphical Identification Criteria — 339
- Adjustment Criterion and Identification — 340
- Workflow #2: Effect Estimation with Bayesian Networks — 344
- Creating a Causal Bayesian Network — 344
- Path Analysis — 349

Pearl's Graph Surgery	352
Introduction to Matching	355
Jouffe's Likelihood Matching	358
Direct Effects Analysis	360

Bibliography   367

Index   371

Preface

While Bayesian networks have flourished in academia over the past three decades, their application for research has developed more slowly. One of the reasons has been the sheer difficulty of generating Bayesian networks for practical research and analytics use. For many years, researchers had to create their own software to utilize Bayesian networks. Needless to say, this made Bayesian networks inaccessible to the vast majority of scientists.

The launch of BayesiaLab 1.0 in 2002 was a major initiative by a newly-formed French company to address this challenge. The development team, lead by Dr. Lionel Jouffe and Dr. Paul Munteanu, designed BayesiaLab with research practitioners in mind—rather than fellow computer scientists. First and foremost, practitioner orientation is reflected in the graphical user interface of BayesiaLab, which allows researchers to work interactively with Bayesian networks in their native form using graphs, as opposed to working with computer code. At the time of writing, BayesiaLab is approaching its sixth major release and has developed into a software platform that provides a comprehensive "laboratory" environment for many research questions.

However, the point-and-click convenience of BayesiaLab does not relieve one of the duty of understanding the fundamentals of Bayesian networks for conducting sound research. With BayesiaLab making Bayesian networks accessible to a much broader audience than ever, demand for the corresponding training has grown tremendously. We recognized the need for a book that supports a self-guided exploration of this field. The objective of this book is to provide a practice-oriented introduction to both Bayesian networks and BayesiaLab.

This book reflects the inherently visual nature of Bayesian networks. Hundreds of illustrations and screenshots provide a tutorial-style explanations of BayesiaLab's core functions. Particularly important steps are repeatedly shown in the context of different examples. The key objective is to provide the reader with step-by-step instructions for transitioning from Bayesian network theory to fully-functional network implementations in BayesiaLab.

The fundamentals of the Bayesian network formalism are linked to numerous disciplines, including computer science, probability theory, information theory, logic, machine learning, and statistics. Also, in terms of applications, Bayesian networks can be utilized in virtually all disciplines. Hence, we meander across many fields of study with the examples presented in this book. Ultimately, we will show how all of them relate to the Bayesian network paradigm. At the same time, we present BayesiaLab as the technology platform, allowing the reader to move immediately from theory to practice. Our goal is to use practical examples for revealing the Bayesian network theory and simultaneously teaching the BayesiaLab technology.

Structure of the Book

Part 1

The intention of the three short chapters in Part 1 of the book is providing a basic familiarity with Bayesian networks and BayesiaLab, from where the reader should feel comfortable to jump into any of the subsequent chapters. For a more cursory observer of this field, Part 1 could serve as an executive summary.

- Chapter 1 provides a motivation for using Bayesian networks from the perspective of analytical modeling.
- Chapter 2 is adapted from Pearl (2000) and introduces the Bayesian network formalism and semantics.
- Chapter 3 presents a brief overview of the BayesiaLab software platform and its core functions.

Part 2

The chapters in Part 2 are mostly self-contained tutorials, which can be studied out of sequence. However, beyond Chapter 8, we assume a certain degree of familiarity with BayesiaLab's core functions.

- In Chapter 4, we discuss how to encode causal knowledge in a Bayesian network for subsequent probabilistic reasoning. In fact, this is the field in which Bayesian networks gained prominence in the 1980s, in the context of building expert systems.
- Chapter 5 introduces data and information theory as a foundation for subsequent chapters. In this context, BayesiaLab's data handling techniques

are presented, such as the **Data Import Wizard**, including **Discretization**. Furthermore, we describe a number of information-theoretic measures that will subsequently be required for machine learning and network analysis.
- Chapter 6 introduces BayesiaLab's **Supervised Learning** algorithms for predictive modeling in the context of a classification task in the field of cancer diagnostics.
- Chapter 7 demonstrates BayesiaLab's **Unsupervised Learning** algorithms for knowledge discovery from financial data.
- Chapter 8 builds on these machine-learning methods and shows a prototypical research workflow for creating a Probabilistic Structural Equation Model for a market research application.
- Chapter 9 deals with missing values, which are typically not of principal research interest but do adversely affect most studies. BayesiaLab leverages conceptual advantages of machine learning and Bayesian networks for reliably imputing missing values.
- Chapter 10 closes the loop by returning to the topic of causality, which we first introduced in Chapter 4. We examine approaches for identifying and estimating causal effects from observational data. Simpson's Paradox serves as the example for this study.

Notation

To clearly distinguish between natural language, software-specific functions, and example-specific jargon, we use the following notation:
- BayesiaLab-specific functions, keywords, commands, and menu items are capitalized and shown in **bold** type. Very frequently used terms, such as "node" or "state" are excluded from this rule in order not to clutter the presentation.
- Names of attributes, variables, node names, node states, and node values are *italicized*.

All highlighted BayesiaLab keywords can also be found in the index.

Chapter 1

1. Introduction

With Professor Judea Pearl receiving the prestigious 2011 A.M. Turing Award, Bayesian networks have presumably received more public recognition than ever before. Judea Pearl's achievement of establishing Bayesian networks as a new paradigm is fittingly summarized by Stuart Russell (2011):

> "[Judea Pearl] is credited with the invention of Bayesian networks, a mathematical formalism for defining complex probability models, as well as the principal algorithms used for inference in these models. This work not only revolutionized the field of artificial intelligence but also became an important tool for many other branches of engineering and the natural sciences. He later created a mathematical framework for causal inference that has had significant impact in the social sciences."

While their theoretical properties made Bayesian networks immediately attractive for academic research, notably concerning the study of causality, only the arrival of practical machine learning algorithms has allowed Bayesian networks to grow beyond their origin in the field of computer science. With the first release of the BayesiaLab software package in 2002, Bayesian networks finally became accessible to a wide range of scientists for use in other disciplines.

All Roads Lead to Bayesian Networks

There are numerous ways we could take to provide motivation for using Bayesian networks. A selection of quotes illustrates that we could approach Bayesian networks from many different perspectives, such as machine learning, probability theory, or knowledge management.

> *"Bayesian networks are as important to AI and machine learning as Boolean circuits are to computer science."* (Stuart Russell in Darwiche, 2009)

> *"Bayesian networks are to probability calculus what spreadsheets are for arithmetic."* (Conrady and Jouffe, 2015)

> *"Currently, Bayesian Networks have become one of the most complete, self-sustained and coherent formalisms used for knowledge acquisition, representation and application through computer systems."* (Bouhamed, 2015)

In this first chapter, however, we approach Bayesian networks from the viewpoint of analytical modeling. Given today's enormous interest in analytics, we wish to relate Bayesian networks to traditional analytic methods from the field of statistics and, furthermore, compare them to more recent innovations in data mining. This context is particularly important given the attention that Big Data and related technologies receive these days. Their dominance in terms of publicity does perhaps drown out some other important methods of scientific inquiry, whose relevance becomes evident by employing Bayesian networks.

Once we have established how Bayesian networks fit into the "world of analytics," Chapter 2 explains the mathematical formalism that underpins the Bayesian network paradigm. For an authoritative account, Chapter 2 is largely based on a technical report by Judea Pearl. While employing Bayesian networks for research has become remarkably easy with BayesiaLab, we need to emphasize the importance of theory. Only a solid understanding of this theory will allow researchers to employ Bayesian networks correctly.

Finally, Chapter 3 concludes the first part of this book with an overview of the BayesiaLab software platform. We show how the theoretical properties of Bayesian networks translate into an capable research tool for many fields of study, ranging from bioinformatics to marketing science and beyond.

Chapter 1

A Map of Analytic Modeling

Following the ideas of Breiman (2001) and Shmueli (2010), we create a map of analytic modeling that is defined by two axes (Figure 1.1):

- The x-axis reflects the **Modeling Purpose**, ranging from **Association/Correlation** to **Causation**. Labels on the x-axis furthermore indicate a conceptual progression, which includes **Description**, **Prediction**, **Explanation**, **Simulation**, and **Optimization**.
- The y-axis represents **Model Source**, i.e. the source of the model specification. **Model Source** ranges from **Theory** (bottom) to **Data** (top). **Theory** is also tagged with **Parametric** as the predominant modeling approach. Additionally, it is tagged with **Human Intelligence**, hinting at the origin of **Theory**. On the opposite end of the y-axis, **Data** is associated with **Machine Learning** and **Artificial Intelligence**. It is also tagged with **Algorithmic** as a contrast to **Parametric** modeling.

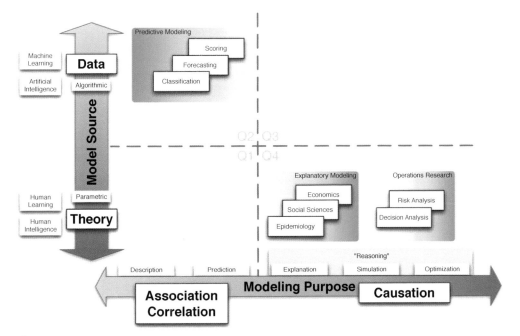

Figure 1.1

Needless to say, Figure 1.1 displays an highly simplified view of the world of analytics, and readers can rightfully point out the limitations of this presentation. Despite this caveat, we will use this map and its coordinate system to position different modeling approaches.

Quadrant 2: Predictive Modeling

Many of today's predictive modeling techniques are algorithmic and would fall mostly into Quadrant 2. In Quadrant 2, a researcher would be primarily interested in the predictive performance of a model, i.e. Y is of interest.

$$\underset{of\ interest}{Y} = f(X) \tag{1.1}$$

Neural networks are a typical example of implementing machine learning techniques in this context. Such models often lack theory. However, they can be excellent "statistical devices" for producing predictions.

Quadrant 4: Explanatory Modeling

In Quadrant 4, the researcher is interested in identifying a model structure that best reflects the underlying "true" data generating process, i.e. we are looking for an explanatory model. Thus, the function f is of greater interest than Y:

$$Y = \underset{of\ interest}{f}(X) \tag{1.2}$$

Traditional statistical techniques that have an explanatory purpose, and which are used in epidemiology and the social sciences, would mostly belong in Quadrant 4. Regressions are the best-known models in this context. Extending further into the causal direction, we would progress into the field of operations research, including simulation and optimization.

Despite the diverging objectives of predictive modeling versus explanatory modeling, i.e. predicting Y versus understanding f, the respective methods are not necessarily incompatible. In Figure 1.1, this is suggested by the blue boxes that gradually fade out as they cross the boundaries and extend beyond their "home" quadrant. However, the best-performing modeling approaches do rarely serve predictive and explanatory purposes equally well. In many situations, the optimal fit-for-purpose models remain very distinct from each other. In fact, Shmueli (2010) has shown that a structurally "less true" model can yield better predictive performance than the "true" explanatory model.

We should also point out that recent advances in machine learning and data mining have mostly occurred in Quadrant 2 and disproportionately benefited predictive modeling. Unfortunately, most machine-learned models are remarkably difficult to interpret in terms of their structural meaning, so new theories are rarely generated this way. For instance, the well-known Netflix Prize competition produced well-performing predictive models, but they yielded little explanatory insight into the structural drivers of choice behavior.

Conversely, in Quadrant 4, deliberately machine learning explanatory models remains rather difficult. As opposed to Quadrant 2, the availability of ever-increasing amounts of data is not necessarily an advantage for discovering theory through machine learning.

Bayesian Networks: Theory and Data

Concerning the horizontal division between **Theory** and **Data** on the **Model Source** axis, Bayesian networks have a special characteristic. Bayesian networks can be built from human knowledge, i.e. from **Theory**, or they can be machine-learned from **Data**. Thus, they can use the entire spectrum as **Model Source** .

Also, due to their graphical structure, machine-learned Bayesian networks are visually interpretable, therefore promoting human learning and theory building. As indicated by the bi-directional arc in Figure 1.2, Bayesian networks allow human learning and machine learning to work in tandem, i.e. Bayesian networks can be developed from a combination of human and artificial intelligence.

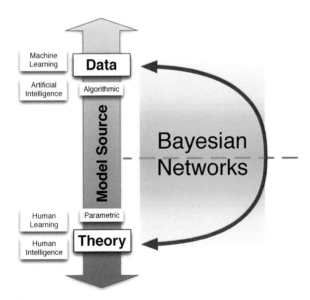

Figure 1.2

Bayesian Networks: Association and Causation

Beyond crossing the boundaries between **Theory** and **Data**, Bayesian networks also have special qualities concerning causality. Under certain conditions and with specific theory-driven assumptions, Bayesian networks facilitate causal inference. In fact, Bayesian network models can cover the entire range from **Association/Correlation** to **Causation**, spanning the entire x-axis of our map (Figure 1.3). In practice, this means that we can add causal assumptions to an existing non-causal network and, thus, create a causal Bayesian network. This is of particular importance when we try to simulate an intervention in a domain, such as estimating the effects of a treatment. In this context, it is imperative to work with a causal model, and Bayesian networks help us make that transition.

▶ Chapter 10. Causal Identification & Estimation, p. 325.

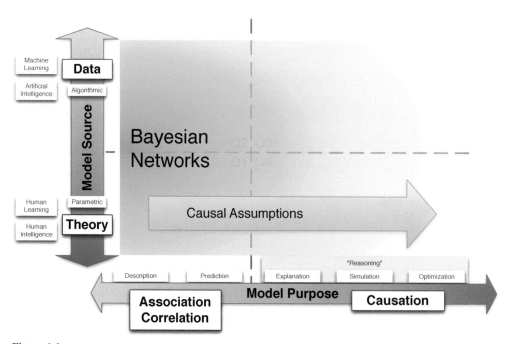

Figure 1.3

As a result, Bayesian networks are a versatile modeling framework, making them suitable for many problem domains. The mathematical formalism underpinning the Bayesian network paradigm will be presented in the next chapter.

Chapter 1

2. Bayesian Network Theory[1]

Probabilistic models based on directed acyclic graphs (DAG) have a long and rich tradition, beginning with the work of geneticist Sewall Wright in the 1920s. Variants have appeared in many fields. Within statistics, such models are known as directed graphical models; within cognitive science and artificial intelligence, such models are known as Bayesian networks. The name honors the Rev. Thomas Bayes (1702-1761), whose rule for updating probabilities in the light of new evidence is the foundation of the approach.

Rev. Bayes addressed both the case of discrete probability distributions of data and the more complicated case of continuous probability distributions. In the discrete case, Bayes' theorem relates the conditional and marginal probabilities of events A and B, provided that the probability of B not equal zero:

$$P(A \mid B) = P(A) \times \frac{P(B \mid A)}{P(B)} \tag{2.1}$$

In Bayes' theorem, each probability has a conventional name: $P(A)$ is the prior probability (or "unconditional" or "marginal" probability) of A. It is "prior" in the sense that it does not take into account any information about B; however, the event B need not occur after event A. In the nineteenth century, the unconditional probability $P(A)$ in Bayes' rule was called the "antecedent" probability; in deductive logic, the antecedent set of propositions and the inference rule imply consequences. The unconditional probability $P(A)$ was called "a priori" by Ronald A. Fisher.

---

[1] This chapter is largely based on Pearl and Russell (2000) and was adapted with permission.

- *P(A|B)* is the conditional probability of *A*, given *B*. It is also called the posterior probability because it is derived from or depends upon the specified value of *B*.
- *P(B|A)* is the conditional probability of *B* given *A*. It is also called the likelihood.
- *P(B)* is the prior or marginal probability of *B*, and acts as a normalizing constant.
- $\frac{P(B|A)}{P(B)}$ is the Bayes factor or likelihood ratio.

Bayes theorem in this form gives a mathematical representation of how the conditional probability of event *A* given *B* is related to the converse conditional probability of *B* given *A*.

The initial development of Bayesian networks in the late 1970s was motivated by the necessity of modeling top-down (semantic) and bottom-up (perceptual) combinations of evidence for inference. The capability for bi-directional inferences, combined with a rigorous probabilistic foundation, led to the rapid emergence of Bayesian networks. They became the method of choice for uncertain reasoning in artificial intelligence and expert systems, replacing earlier, ad hoc rule-based schemes.

Bayesian networks are models that consist of two parts, a qualitative one based on a DAG for indicating the dependencies, and a quantitative one based on local probability distributions for specifying the probabilistic relationships. The DAG consists of nodes and directed links:

- Nodes represent variables of interest (e.g. the temperature of a device, the gender of a patient, a feature of an object, the occurrence of an event). Even though Bayesian networks can handle continuous variables, we exclusively discuss Bayesian networks with discrete nodes in this book. Such nodes can correspond to symbolic/categorical variables, numerical variables with discrete values, or discretized continuous variables.
- Directed links represent statistical (informational) or causal dependencies among the variables. The directions are used to define kinship relations, i.e. parent-child relationships. For example, in a Bayesian network with a link from *X* to *Y*, *X* is the parent node of *Y*, and *Y* is the child node.

▸ Chapter 10. Causal Identification & Estimation, p. 325.

The local probability distributions can be either marginal, for nodes *without* parents (root nodes), or *conditional*, for nodes with parents. In the latter case, the dependencies are quantified by conditional probability tables (CPT) for each node given its parents in the graph.

Chapter 2

Once fully specified, a Bayesian network compactly represents the joint probability distribution (JPD) and, thus, can be used for computing the posterior probabilities of any subset of variables given evidence[2] about any other subset.

A Non-Causal Bayesian Network Example

Figure 2.1 shows a simple Bayesian network, which consists of only two nodes and one link. It represents the JPD of the variables *Eye Color* and *Hair Color* in a population of students (Snee, 1974). In this case, the conditional probabilities of *Hair Color* given the values of its parent node, *Eye Color*, are provided in a CPT. It is important to point out that this Bayesian network does not contain any causal assumptions, i.e. we have no knowledge of the causal order between the variables. Thus, the interpretation of this network should be merely statistical (informational).

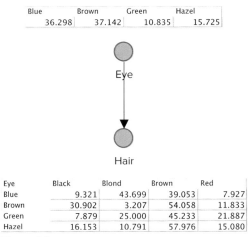

Figure 2.1

A Causal Network Example

Figure 2.2 illustrates another simple yet typical Bayesian network. In contrast to the statistical relationships in Figure 2.1, the diagram in Figure 2.2 describes the causal relationships among the seasons of the year (X_1), whether it is raining (X_2), whether the sprinkler is on (X_3), whether the pavement is wet (X_4), and whether the pavement is slippery (X_5). Here, the absence of a direct link between X_1 and X_5, for example, captures our understanding that there is no direct influence of season on slipperiness.

---

2   Throughout this book we use "setting evidence on a variable" and "observing a variable" interchangeably.

The influence is mediated by the wetness of the pavement (if freezing were a possibility, a direct link could be added).

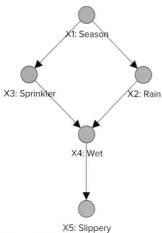

Figure 2.2

Perhaps the most important aspect of Bayesian networks is that they are direct representations of the world, not of reasoning processes. The arrows in the diagram represent real causal connections and not the flow of information during reasoning (as in rule-based systems and neural networks). Reasoning processes can operate on Bayesian networks by propagating information in any direction. For example, if the sprinkler is on, then the pavement is probably wet (prediction, simulation). If someone slips on the pavement, that will also provide evidence that it is wet (abduction, reasoning to a probable cause, or diagnosis). On the other hand, if we see that the pavement is wet, that will make it more likely that the sprinkler is on or that it is raining (abduction); but if we then observe that the sprinkler is on, that will reduce the likelihood that it is raining (explaining away). It is the latter form of reasoning, explaining away, that is especially difficult to model in rule-based systems and neural networks in a natural way, because it seems to require the propagation of information in two directions.

A Dynamic Bayesian Network Example

Entities that live in a changing environment must keep track of variables whose values change over time. Dynamic Bayesian networks capture this process by representing multiple copies of the state variables, one for each time step. A set of variables X_{t-1} and X_t denotes the world state at times *t-1* and *t* respectively. A set of evidence variables E_t denotes the observations available at time *t*. The sensor model $P(E_t|X_t)$ is encoded in the conditional probability distributions for the observable variables, given the state variables. The transition model $P(X_t|X_{t-1})$ relates the state at time $t-1$ to the state at time t. Keeping track of the world means computing the current probability distribution over world states given all past observations, i.e. $P(X_t|E_1,...,E_t)$.

Dynamic Bayesian networks (DBN) are a generalization of Hidden Markov Models (HMM) and Kalman Filters (KF). Every HMM and KF can be represented with a DBN. Furthermore, the DBN representation of an HMM is much more compact and, thus, much better understandable. The nodes in the HMM represent the states of the system, whereas the nodes in the DBN represent the dimensions of the system. For example, the HMM representation of the valve system in Figure 2.3 is made of 26 nodes and 36 arcs, versus 9 nodes and 11 arcs in the DBN (Weber and Jouffe, 2003).

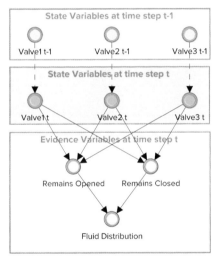

Figure 2.3

Representation of the Joint Probability Distribution

Any complete probabilistic model of a domain must—either explicitly or implicitly—represent the joint probability distribution (JPD), i.e. the probability of every possible event as defined by the combination of the values of all the variables. There are exponentially many such events, yet Bayesian networks achieve compactness by factoring the JPD into local, conditional distributions for each variable given its parents. If x_i denotes some value of the variable X_i and pa_i denotes some set of values for the parents of X_i, then $P(x_i|pa_i)$ denotes this conditional probability distribution. For example, in the graph in Figure 2.4, $P(x_4|x_2,x_3)$ is the probability of *Wetness* given the values of *Sprinkler* and *Rain*. The global semantics of Bayesian networks specifies that the full JPD is given by the product rule (or chain rule):

$$P(x_i,...,x_n) = \prod_i P(x_i \mid pa_i) \tag{2.2}$$

In our example network, we have:

$$P(x_1,x_2,x_3,x_4,x_5) = P(x_1)P(x_2 \mid x_1)P(x_3 \mid x_1)P(x_4 \mid x_2,x_3)P(x_5 \mid x_4) \tag{2.3}$$

It becomes clear that the number of parameters grows linearly with the size of the network, i.e. the number of variables, whereas the size of the JPD itself grows exponentially. Given a discrete representation of the CPD with a CPT, the size of a local CPD grows exponentially with the number of parents. Savings can be achieved using compact CPD representations—such as noisy-OR models, trees, or neural networks.

The JPD representation with Bayesian networks also translates into a local semantics, which asserts that each variable is independent of non-descendants in the network given its parents. For example, the parents of X_4 in Figure 2.4 are X_2 and X_3, and they render X_4 independent of the remaining non-descendant, X_1:

$$P(x_4 \mid x_1,x_2,x_3) = P(x_4 \mid x_2,x_3) \tag{2.4}$$

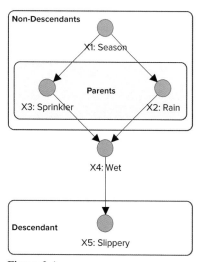

Figure 2.4

The collection of independence assertions formed in this way suffices to derive the global assertion of the product rule (or chain rule) in (2.2), and vice versa. The local semantics is most useful for constructing Bayesian networks because selecting as parents all the direct causes (or direct relationships) of a given variable invariably satisfies the local conditional independence conditions. The global semantics leads directly to a variety of algorithms for reasoning.

Evidential Reasoning

From the product rule (or chain rule) in (2.2), one can express the probability of any desired proposition in terms of the conditional probabilities specified in the network. For example, the probability that the *Sprinkler* is on given that the *Pavement* is slippery is:

$$P(X_3 = on \mid X_5 = true) = \frac{P(X_3 = on, X_5 = true)}{P(X_5 = true)} \qquad (2.5)$$

$$= \frac{\sum_{x_1,x_2,x_4} P(x_1, x_2, X_3 = on, x_4, X_5 = true)}{\sum_{x_1,x_2,x_3,x_4} P(x_1, x_2, x_3, x_4, X_5 = true)}$$

$$= \frac{\sum_{x_1,x_2,x_4} P(x_1)(x_2 \mid x_1) P(X_3 = on \mid x_1) P(x_4 \mid x_2, X_3 = on) P(X_5 = true \mid x_4)}{\sum_{x_1,x_2,x_3,x_4} P(x_1) P(x_2 \mid x_1) P(x_3 \mid x_1) P(x_4 \mid x_2, x_3) P(X_5 = true \mid x_4)}$$

These expressions can often be simplified in ways that reflect the structure of the network itself. The first algorithms proposed for probabilistic calculations in Bayesian networks used a local distributed message-passing architecture, typical of many cognitive activities. Initially, this approach was limited to tree-structured networks but was later extended to general networks in Lauritzen and Spiegelhalter's (1988) method of junction tree propagation. A number of other exact methods have been developed and can be found in recent textbooks.

It is easy to show that reasoning in Bayesian networks subsumes the satisfiability problem in propositional logic and, therefore, exact inference is NP-hard. Monte Carlo simulation methods can be used for approximate inference (Pearl, 1988) giving gradually improving estimates as sampling proceeds. These methods use local message propagation on the original network structure, unlike junction-tree methods. Alternatively, variational methods provide bounds on the true probability.

Causal Reasoning

Most probabilistic models, including general Bayesian networks, describe a joint probability distribution (JPD) over possible observed events, but say nothing about what will happen if a certain intervention occurs. For example, what if I turn the

Sprinkler on instead of just observing that it is turned on? What effect does that have on the *Season*, or on the connection between *Wet* and *Slippery*? A causal network, intuitively speaking, is a Bayesian network with the added property that the parents of each node are its direct causes, as in Figure 2.4. In such a network, the result of an intervention is obvious: the *Sprinkler* node is set to $X_3=on$ and the causal link between the *Season* X_1 and the *Sprinkler* X_3 is removed (Figure 2.5). All other causal links and conditional probabilities remain intact, so the new model is:

$$P(x_1,x_2,x_3,x_4) = P(x_1)P(x_2 \mid x_1)P(x_4 \mid x_2, X_3 = on)P(x_5 \mid x_4) \qquad (2.6)$$

Notice that this differs from observing that $X_3=on$, which would result in a new model that included the term $P(X_3=on|x_1)$. This mirrors the difference between seeing and doing: after observing that the *Sprinkler* is on, we wish to infer that the *Season* is dry, that it probably did not rain, and so on. An arbitrary decision to turn on the *Sprinkler* should not result in any such beliefs.

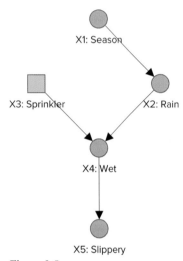

Figure 2.5

Causal networks are more properly defined, then, as Bayesian networks in which the correct probability model—after intervening to fix any node's value—is given simply by deleting links from the node's parents. For example, *Fire → Smoke* is a causal network, whereas *Smoke → Fire* is not, even though both networks are equally capable of representing any joint probability distribution of the two variables. Causal networks model the environment as a collection of stable component mechanisms. These mechanisms may be reconfigured locally by interventions, with corresponding local changes in the model. This, in turn, allows causal networks to be used very naturally for prediction by an agent that is considering various courses of action.

Chapter 2

Learning Bayesian Network Parameters

Given a qualitative Bayesian network structure, the conditional probability tables, $P(x_i|pa_i)$, are typically estimated with the maximum likelihood approach from the observed frequencies in the dataset associated with the network.

In pure Bayesian approaches, Bayesian networks are designed from expert knowledge and include hyperparameter nodes. Data (usually scarce) is used as pieces of evidence for incrementally updating the distributions of the hyperparameters (**Bayesian Updating**).

▸ Bayesian Updating in Chapter 3, p. 38.

Learning Bayesian Network Structure

It is also possible to machine learn the structure of a Bayesian network, and two families of methods are available for that purpose. The first one, using constraint-based algorithms, is based on the probabilistic semantic of Bayesian networks. Links are added or deleted according to the results of statistical tests, which identify marginal and conditional independencies. The second approach, using score-based algorithms, is based on a metric that measures the quality of candidate networks with respect to the observed data. This metric trades off network complexity against the degree of fit to the data, which is typically expressed as the likelihood of the data given the network.

As a substrate for learning, Bayesian networks have the advantage that it is relatively easy to encode prior knowledge in network form, either by fixing portions of the structure, forbidding relations, or by using prior distributions over the network parameters. Such prior knowledge can allow a system to learn accurate models from much fewer data than are required for clean sheet approaches.

Causal Discovery

One of the most exciting prospects in recent years has been the possibility of using Bayesian networks to discover causal structures in raw statistical data—a task previously considered impossible without controlled experiments. Consider, for example, the following intransitive pattern of dependencies among three events: A and B are dependent, B and C are dependent, yet A and C are independent. If you asked a person to supply an example of three such events, the example would invariably portray A and C as two independent causes and B as their common effect, namely $A \rightarrow B \leftarrow C$.

For instance, *A* and *C* could be the outcomes of two fair coins, and *B* represents a bell that rings whenever either coin comes up heads.

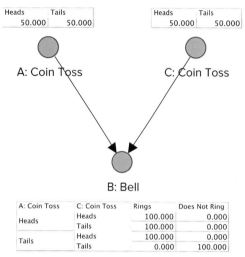

Figure 2.6

Fitting this dependence pattern with a scenario in which *B* is the cause and *A* and *C* are the effects is mathematically feasible but very unnatural, because it must entail fine tuning of the probabilities involved. The desired dependence pattern will be destroyed as soon as the probabilities undergo a slight change.

Such thought experiments tell us that certain patterns of dependency, which are totally void of temporal information, are conceptually characteristic of certain causal directionalities and not others. When put together systematically, such patterns can be used to infer causal structures from raw data and to guarantee that any alternative structure compatible with the data must be less stable than the one(s) inferred; namely slight fluctuations in parameters will render that structure incompatible with the data.

Caveat

Despite recent advances, causal discovery is an area of active research, with countless questions remaining unresolved. Thus, no generally accepted causal discovery algorithms are currently available for applied researchers. As a result, all causal networks presented in this book are constructed from expert knowledge, or machine-learned and then validated as causal by experts. The assumptions necessary for a causal interpretation of a Bayesian network will be discussed in Chapter 10.

Chapter 2

Chapter 3

3. BayesiaLab

While the conceptual advantages of Bayesian networks had been known in the world of academia for some time, leveraging these properties for practical research applications was very difficult for non-computer scientists prior to BayesiaLab's first release in 2002.

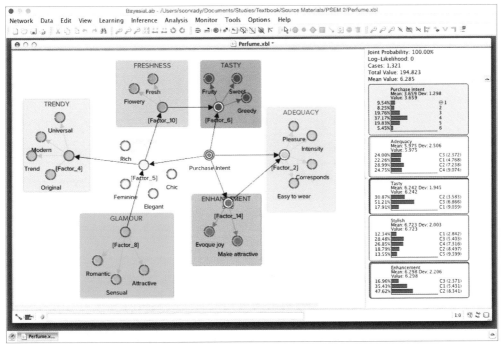

Figure 3.1

BayesiaLab is a powerful desktop application (Windows/Mac/Unix) with a sophisticated graphical user interface, which provides scientists a comprehensive "laboratory" environment for machine learning, knowledge modeling, diagnosis, analysis, simulation, and optimization. With BayesiaLab, Bayesian networks have become practical for gaining deep insights into problem domains. BayesiaLab leverages the inherently graphical structure of Bayesian networks for exploring and explaining complex problems. Figure 3.1 shows a screenshot of a typical research project.

BayesiaLab is the result of nearly twenty years of research and software development by Dr. Lionel Jouffe and Dr. Paul Munteanu. In 2001, their research efforts led to the formation of Bayesia S.A.S., headquartered in Laval in northwestern France. Today, the company is the world's leading supplier of Bayesian network software, serving hundreds major corporations and research organizations around the world.

BayesiaLab's Methods, Features, and Functions

As conceptualized in the diagram in Figure 3.2, BayesiaLab is designed around a prototypical workflow with a Bayesian network model at the center. BayesiaLab supports the research process from model generation to analysis, simulation, and optimization. The entire process is fully contained in a uniform "lab" environment, which provides scientists with flexibility in moving back and forth between different elements of the research task.

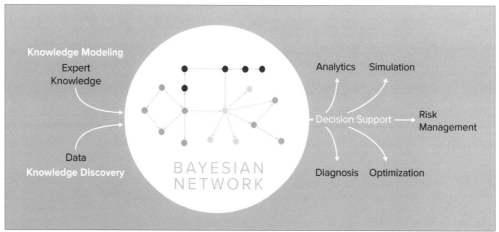

Figure 3.2

In Chapter 1, we presented our principal motivation for using Bayesian networks, namely their universal suitability across the entire "map" of analytic modeling: Bayesian networks can be modeled from pure theory, and they can be learned from data alone; Bayesian networks can serve as predictive models, and they can represent causal relationships. Figure 3.3 shows how our claim of "universal modeling capability" translates into specific functions provided by BayesiaLab, which are placed as blue boxes on the analytics map.

Chapter 3

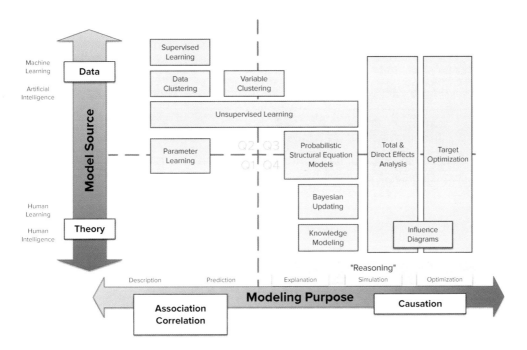

Figure 3.3

Knowledge Modeling

Subject matter experts often express their causal understanding of a domain in the form of diagrams, in which arrows indicate causal directions. This visual representation of causes and effects has a direct analog in the network graph in BayesiaLab. Nodes (representing variables) can be added and positioned on BayesiaLab's **Graph Panel** with a mouse-click, arcs (representing relationships) can be "drawn" between nodes. The causal direction can be encoded by orienting the arcs from cause to effect (Figure 3.4).

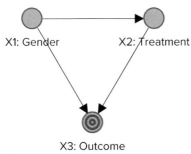

Figure 3.4

The quantitative nature of relationships between variables, plus many other attributes, can be managed in BayesiaLab's **Node Editor**. In this way, BayesiaLab facilitates the straightforward encoding of one's understanding of a domain. Simultaneously, BayesiaLab enforces internal consistency, so that impossible conditions cannot be encoded accidentally. In Chapter 4, we will present a practical example of causal knowledge modeling, followed by probabilistic reasoning.

In addition to having individuals directly encode their explicit knowledge in BayesiaLab, the **Bayesia Expert Knowledge Elicitation Environment (BEKEE)**[1] is available for acquiring the probabilities of a network from a group of experts. BEKEE offers a web-based interface for systematically eliciting explicit and tacit knowledge from multiple stakeholders.

Discrete, Nonlinear and Nonparametric Modeling

BayesiaLab contains all "parameters" describing probabilistic relationships between variables in conditional probability tables (CPT), which means that no functional forms are utilized.[2] Given this nonparametric, discrete approach, BayesiaLab can conveniently handle nonlinear relationships between variables. However, this CPT-based representation requires a preparation step for dealing with continuous variables, namely discretization. This consists in defining—manually or automatically—a discrete representation of all continuous values. BayesiaLab offers several tools for discretization, which are accessible in the **Data Import Wizard**, in the **Node Editor** (Figure 3.5), and in a standalone **Discretization** function. In this context, univariate, bivariate, and multivariate discretization algorithms are available.

---

1 BEKEE is an optional subscription service. See www.bayesia.us/bekee.

2 BayesiaLab can utilize formulas and trees to compactly describe the CPT, however, the internal representation remains table-based.

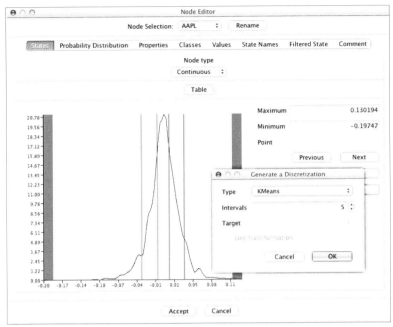

Figure 3.5

Missing Values Processing

BayesiaLab offers a range of sophisticated methods for missing values processing. During network learning, BayesiaLab performs missing values processing automatically "behind the scenes". More specifically, the **Structural EM** algorithm or the **Dynamic Imputation** algorithms are applied after each modification of the network during learning, i.e. after every single arc addition, suppression and inversion. Bayesian networks provide a few fundamental advantages for dealing with missing values. In Chapter 9, we will focus exclusively on this topic.

▸ Chapter 9. Missing Values Processing, p. 289.

Parameter Estimation

Parameter Estimation with BayesiaLab is at the intersection of theory-driven and data-driven modeling. For a network that was generated either from expert knowledge or through machine learning, BayesiaLab can use the observations contained in an associated dataset to populate the CPT via **Maximum Likelihood Estimation**.

Bayesian Updating

In general, Bayesian networks are nonparametric models. However, a Bayesian network can also serve as a parametric model if an expert uses equations for defining local CPDs and, additionally, specifies hyperparameters, i.e. nodes that explicitly represent parameters that are used in the equations.

As opposed to BayesiaLab's usual parameter estimation via **Maximum Likelihood**, the associated dataset provides pieces of evidence for incrementally updating—via probabilistic inference—the distributions of the hyperparameters.

Machine Learning

Despite our repeated emphasis on the relevance of human expert knowledge, especially for identifying causal relations, much of this book is dedicated to acquiring knowledge from data through machine learning. BayesiaLab features a comprehensive array of highly optimized learning algorithms that can quickly uncover structures in datasets. The optimization criteria in BayesiaLab's learning algorithms are based on information theory (e.g. the **Minimum Description Length**). With that, no assumptions regarding the variable distributions are made. These algorithms can be used for all kinds and all sizes of problem domains, sometimes including thousands of variables with millions of potentially relevant relationships.

Unsupervised Structural Learning (Quadrant 2/3)

In statistics, "unsupervised learning" is typically understood to be a classification or clustering task. To make a very clear distinction, we place emphasis on "structural" in "Unsupervised Structural Learning," which covers a number of important algorithms in BayesiaLab.

Unsupervised Structural Learning means that BayesiaLab can discover probabilistic relationships between a large number of variables, without having to specify input or output nodes. One might say that this is a quintessential form of knowledge discovery, as no assumptions are required to perform these algorithms on unknown datasets (Figure 3.6).

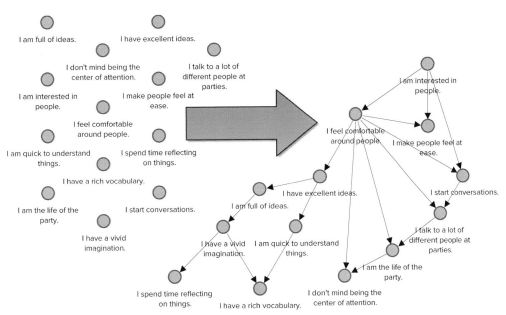

Figure 3.6

Supervised Learning (Quadrant 2)

Supervised Learning in BayesiaLab has the same objective as many traditional modeling methods, i.e. to develop a model for predicting a target variable. Note that numerous statistical packages also offer "Bayesian Networks" as a predictive modeling technique. However, in most cases, these packages are restricted in their capabilities to a one type of network, i.e. the **Naive Bayes** network. BayesiaLab offers a much greater number of **Supervised Learning** algorithms to search for the Bayesian network that best predicts the target variable while also taking into account the complexity of the resulting network (Figure 3.7).

We should highlight the **Markov Blanket** algorithm for its speed, which is particularly helpful when dealing with a large number of variables. In this context, the **Markov Blanket** algorithm can serve as an efficient variable selection algorithm. An example of **Supervised Learning** using this algorithm, and the closely-related **Augmented Markov Blanket** algorithm, will be presented in Chapter 6.

▶ Markov Blanket Definition in Chapter 6, p. 124.

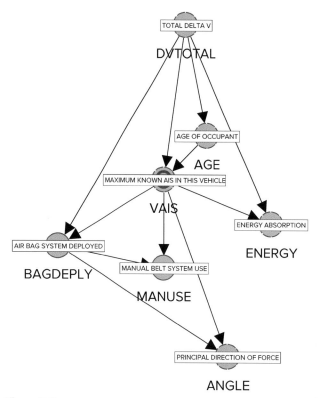

Figure 3.7

Clustering (Quadrant 2/3)

Clustering in BayesiaLab covers both **Data Clustering** and **Variable Clustering**. The former applies to the grouping of records (or observations) in a dataset;[3] the latter performs a grouping of variables according to the strength of their mutual relationships (Figure 3.8).

A third variation of this concept is of particular importance in BayesiaLab: **Multiple Clustering** can be characterized as a kind of nonlinear, nonparametric and nonorthogonal factor analysis. **Multiple Clustering** often serves as the basis for developing **Probabilistic Structural Equation Models** (Quadrant 3/4) with BayesiaLab.

▶ Chapter 8. Probabilistic Structural Equation Models, p. 201.

---

3 Throughout this book, we use "dataset" and "database" interchangeably.

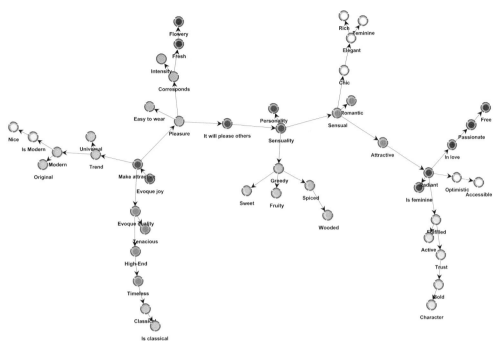

Figure 3.8

Inference: Diagnosis, Prediction, and Simulation

The inherent ability of Bayesian networks to explicitly model uncertainty makes them suitable for a broad range of real-world applications. In the Bayesian network framework, diagnosis, prediction, and simulation are identical computations. They all consist of observational inference conditional upon evidence:

- Inference from effect to cause: diagnosis or abduction.
- Inference from cause to effect: simulation or prediction.

This distinction, however, only exists from the perspective of the researcher, who would presumably see the symptom of a disease as the effect and the disease itself as the cause. Hence, carrying out inference based on observed symptoms is interpreted as "diagnosis."

Observational Inference (Quadrant 1/2)

One of the central benefits of Bayesian networks is that they compute inference "omni-directionally." Given an observation with any type of evidence on any of the networks' nodes (or a subset of nodes), BayesiaLab can compute the posterior probabilities of all other nodes in the network, regardless of arc direction. Both exact and approximate observational inference algorithms are implemented in BayesiaLab. We briefly illustrate evidence-setting and inference with the expert system network shown in Figure 3.9.[4]

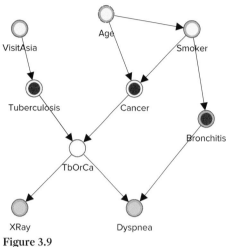

Figure 3.9

Types of Evidence

▸ Inference with Probabilistic and Numerical Evidence in Chapter 7, p. 188.

1. **Hard Evidence**: no uncertainty regarding the state of the variable (node), e.g. *P(Smoker=True)=100%* (Figure 3.10).

2. **Probabilistic Evidence (or Soft Evidence)**, defined by marginal probability distributions: *P(Bronchitis=True)=66.67%* (Figure 3.11).

3. **Numerical Evidence**, for numerical variables, or for categorical/symbolic variables that have associated numerical values. BayesiaLab computes a marginal probability distribution to generate the specified expected value: E(*Age*)=*39* (Figure 3.12).

4. **Likelihood Evidence** (or **Virtual Evidence**), defined by a likelihood of each state, ranging from 0%, i.e. impossible, to 100%, which means that no evidence reduces the probability of the state. To be valid as evidence, the sum of the likelihoods must be greater than 0. Also, note that the upper boundary for the sum of the likelihoods equals the number of states.

---

4 This example is adapted from Lauritzen and Spiegelhalter (1988).

Chapter 3

Setting the same likelihood to all states corresponds to setting no evidence at all (Figure 3.13).

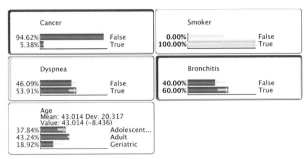

Figure 3.10

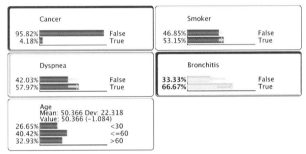

Figure 3.11

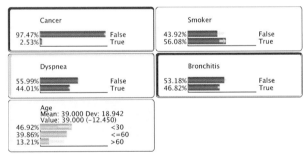

Figure 3.12

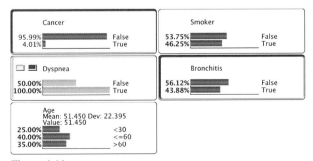

Figure 3.13

43

Causal Inference (Quadrant 3/4)

Beyond observational inference, BayesiaLab can also perform causal inference for computing the impact of *intervening* on a subset of variables instead of merely *observing* these variables. Both Pearl's **Do-Operator** and Jouffe's **Likelihood Matching** are available for this purpose. We will provide a detailed discussion of causal inference in Chapter 10.

Effects Analysis (Quadrants 3/4)

Many research activities focus on estimating the size of an effect, e.g. to establish the treatment effect of a new drug or to determine the sales boost from a new advertising campaign. Other studies attempt to decompose observed effects into their causes, i.e. they perform attribution.

BayesiaLab performs simulations to compute effects, as parameters as such do not exist in this nonparametric framework. As all the dynamics of the domain are encoded in discrete **CPTs**, effect sizes only manifest themselves when different conditions are simulated. **Total Effects Analysis**, **Target Mean Analysis**, and several other functions offer ways to study effects, including nonlinear effects and variables interactions.

Optimization (Quadrant 4)

BayesiaLab's ability to perform inference over all possible states of all nodes in a network also provides the basis for searching for node values that optimize a target criterion. BayesiaLab's **Target Dynamic Profile** and **Target Optimization** are a set of tools for this purpose.

▶ Target Dynamic Profile in Chapter 8, p. 274.

Using these functions in combination with **Direct Effects** is of particular interest when searching for the optimum combination of variables that have a nonlinear relationship with the target, plus co-relations between them. A typical example would be searching for the optimum mix of marketing expenditures to maximize sales. BayesiaLab's **Target Optimization** will search, within the specified constraints, for those scenarios that optimize the target criterion (Figure 3.14). An example of **Target Dynamic Profile** will be presented in Chapter 8.

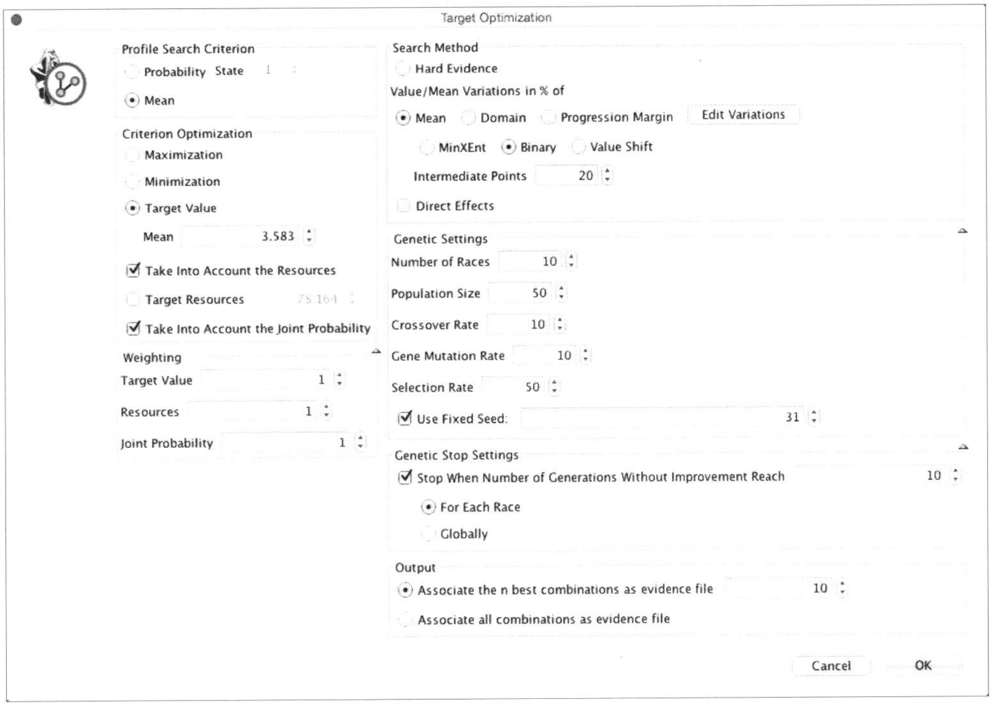

Figure 3.14

Model Utilization

BayesiaLab provides a range of functions for systematically utilizing the knowledge contained in a Bayesian network. They make a network accessible as an expert system that can be queried interactively by an end user or through an automated process.

The **Adaptive Questionnaire** function provides guidance in terms of the optimum sequence for seeking evidence. BayesiaLab determines dynamically, given the evidence already gathered, the next best piece of evidence to obtain, in order to maximize the information gain with respect to the target variable, while minimizing the cost of acquiring such evidence. In a medical context, for instance, this would allow for the optimal "escalation" of diagnostic procedures, from "low-cost/small-gain" evidence (e.g. measuring the patient's blood pressure) to "high-cost/large-gain" evidence (e.g. performing an MRI scan). The **Adaptive Questionnaire** will be presented in the context of an example about tumor classification in Chapter 6.

▸ Adaptive Questionnaire in Chapter 6, p. 147.

The **WebSimulator** is a platform for publishing interactive models and **Adaptive Questionnaires** via the web, which means that any Bayesian network model built with BayesiaLab can be shared privately with clients or publicly with a broader audience. Once a model is published via the **WebSimulator**, end users can try out scenarios and examine the dynamics of that model (Figure 3.15).

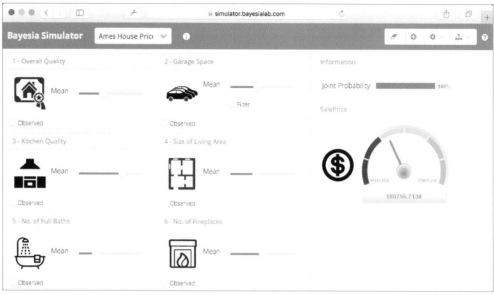

Figure 3.15

Batch Inference is available for automatically performing inference on a large number of records in a dataset. For example, **Batch Inference** can be used to produce a predictive score for all customers in a database. With the same objective, BayesiaLab's optional **Export** function can translate predictive network models into static code that can run in external programs. Modules are available that can generate code for R, SAS, PHP, VBA, and JavaScript.

Developers can also access many of BayesiaLab's functions—outside the graphical user interface—by using the **Bayesia Engine APIs**. The **Bayesia Modeling Engine** allows constructing and editing networks. The **Bayesia Inference Engine** can access network models programmatically for performing automated inference, e.g. as part of a real-time application with streaming data. The **Bayesia Engine APIs** are implemented as pure Java class libraries (jar files), which can be integrated into any software project.

Knowledge Communication

While generating a Bayesian network, either by expert knowledge modeling or through machine learning, is all about a computer acquiring knowledge, a Bayesian network can also be a remarkably powerful tool for humans to extract or "harvest" knowledge. Given that a Bayesian network can serve as a high-dimensional representation of a real-world domain, BayesiaLab allows us to interactively—even playfully—engage with this domain to learn about it (Figure 3.16). Through visualization, simulation, and analysis functions, plus the graphical nature of the network model itself, BayesiaLab becomes an instructional device that can effectively retrieve and communicate the knowledge contained within the Bayesian network. As such, BayesiaLab becomes a bridge between artificial intelligence and human intelligence.

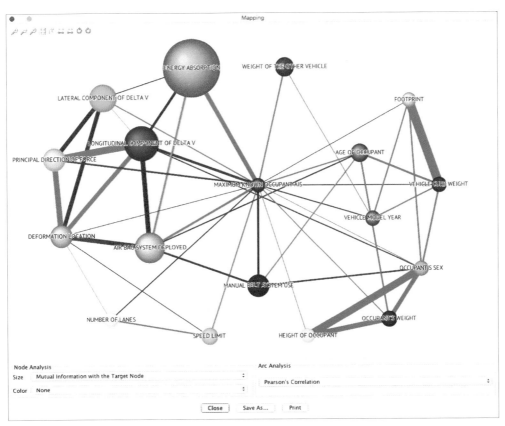

Figure 3.16

Chapter 4

4. Knowledge Modeling & Reasoning

This chapter presents a workflow for encoding expert knowledge and subsequently performing omni-directional probabilistic inference in the context of a real-world reasoning problem. While Chapter 1 provided a general motivation for using Bayesian networks as an analytics framework, this chapter highlights the perhaps unexpected relevance of Bayesian networks for reasoning in everyday life. The example proves that "common-sense" reasoning can be rather tricky. On the other hand, encoding "common-sense knowledge" in a Bayesian network turns out to be uncomplicated. We want to demonstrate that reasoning with Bayesian networks can be as straightforward as doing arithmetic with a spreadsheet.

Background & Motivation

Complexity & Cognitive Challenges

It is presumably fair to state that reasoning in complex environments creates cognitive challenges for humans. Adding uncertainty to our observations of the problem domain, or even considering uncertainty regarding the structure of the domain itself, makes matters worse. When uncertainty blurs so many premises, it can be particularly difficult to find a common reasoning framework for a group of stakeholders.

No Data, No Analytics.

If we had hard observations from our domain in the form of data, it would be quite natural to build a traditional analytic model for decision support. However, the real world often yields only fragmented data or no data at all. It is not uncommon that we merely have the opinions of individuals who are more or less familiar with the problem domain.

To an Analyst With Excel, Every Problem Looks Like Arithmetic.

In the business world, it is typical to use spreadsheets to model the relationships between variables in a problem domain. Also, in the absence of hard observations, it is reasonable that experts provide assumptions instead of data. Any such expert knowledge is typically encoded in the form of single-point estimates and formulas. However, using of single values and formulas instantly oversimplifies the problem domain: firstly, the variables, and the relationships between them, become deterministic; secondly, the left-hand side versus right-hand side nature of formulas restricts inference to only one direction.

Taking No Chances!

Given that cells and formulas in spreadsheets are deterministic and only work with single-point values, they are well suited for encoding "hard" logic, but not at all for "soft" probabilistic knowledge that includes uncertainty. As a result, any uncertainty has to be addressed with workarounds, often in the form of trying out multiple scenarios or by working with simulation add-ons.

It Is a One-Way Street!

The lack of omni-directional inference, however, may the bigger issue in spreadsheets. As soon as we create a formula linking two cells in a spreadsheet, e.g. *B1=function(A1)*, we preclude any evaluation in the opposite direction, from *B1* to *A1*.

Assuming that *A1* is the cause, and *B1* is the effect, we can indeed use a spreadsheet for inference in the causal direction, i.e. perform a simulation. However, even if we were certain about the causal direction between them, unidirectionality would remain a concern. For instance, if we were only able to observe the effect *B1*, we could not infer the cause *A1*, i.e. we could not perform a diagnosis from effect to cause. The one-way nature of spreadsheet computations prevents this.

Bayesian Networks to the Rescue!

Bayesian networks are probabilistic by default and handle uncertainty "natively." A Bayesian network model can work directly with probabilistic inputs, probabilistic relationships, and deliver correctly computed probabilistic outputs. Also, whereas traditional models and spreadsheets are of the form *y=f(x)*, Bayesian networks do not have to distinguish between independent and dependent variables. Rather, a

Bayesian network represents the entire joint probability distribution of the system under study. This representation facilitates omni-directional inference, which is what we typically require for reasoning about a complex problem domain, such as the example in this chapter.

Example: Where is My Bag?

While most other examples in this book resemble proper research topics, we present a rather casual narrative to introduce probabilistic reasoning with Bayesian networks. It is a common situation taken straight from daily life, for which a "common-sense interpretation" may appear more natural than our proposed formal approach. As we shall see, dealing formally with informal knowledge provides a robust basis for reasoning under uncertainty.

Did My Checked Luggage Make the Connection?

Most travelers will be familiar with the following hypothetical situation, or something fairly similar: You are traveling between two cities and need to make a flight connection in a major hub. Your first flight segment (from the origin city to the hub) is significantly delayed, and you arrive at the hub with barely enough time to make the connection. The boarding process is already underway by the time you get to the departure gate of your second flight segment (from the hub to the final destination).

Problem #1

Out of breath, you check in with the gate agent, who informs you that the luggage you checked at the origin airport may or may not make the connection. She states apologetically that there is only a 50/50 chance that you will get your bag upon arrival at your destination airport.

Once you have landed at your destination airport, you head straight to baggage claim and wait for the first pieces of luggage to appear on the baggage carousel. Bags come down the chute onto the carousel at a steady rate. After five minutes of watching fellow travelers retrieve their luggage, you wonder what the chances are that you will ultimately get your bag. You reason that if the bag had indeed made it onto the plane, it would be increasingly likely for it to appear among the remaining pieces to be unloaded. However, you do not know for sure that your piece was actually on the plane. Then, you think, you better get in line to file a claim at the baggage office. Is that reasonable? As you wait, how should you update your expectation about getting your bag?

Problem #2

Just as you contemplate your next move, you see a colleague picking up his suitcase. As it turns out, your colleague was traveling on the very same itinerary as you. His luggage obviously made it, so you conclude that you better wait at the carousel for the very last piece to be delivered. How does the observation of your colleague's suitcase change your belief in the arrival of your bag? Does all that even matter? After all, the bag either made the connection or not. The fact that you now observe something after the fact cannot influence what happened earlier, right?

Knowledge Modeling for Problem #1

This problem domain can be explained by a causal Bayesian network, only using a few common-sense assumptions. We demonstrate how we can combine different pieces of available —but uncertain—knowledge into a network model. Our objective is to calculate the correct degree of belief in the arrival of your luggage as a function of time and your own observations.

As per our narrative, we obtain the first piece of information from the gate agent. She says that there is a 50/50 chance that your bag is on the plane. More formally, we express this as:

$$P(\textit{Your Bag on Plane} = \textit{True}) = 0.5 \tag{4.1}$$

We encode this probabilistic knowledge in a Bayesian network by creating a node. In BayesiaLab, we click the **Node Creation** icon (●) and then point to the desired position on the **Graph Panel**.

Chapter 4

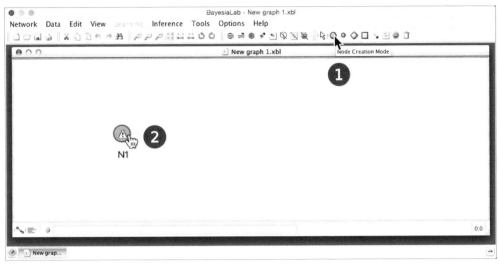

Figure 4.1

Once the node is in place, we update its name to "Your Bag on Plane" by double-clicking the default name *N1*. Then, by double-clicking the node itself, we open up BayesiaLab's **Node Editor**. Under the tab **Probability Distribution > Probabilistic**, we define the probability that *Your Bag on Plane=True*, which is 50%, as per the gate agent's statement. Given that these probabilities do not depend on any other variables, we speak of marginal probabilities (Figure 4.2). Note that in BayesiaLab probabilities are always expressed as percentages.

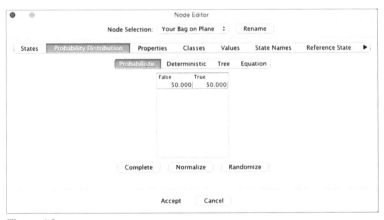

Figure 4.2

Assuming that there is no other opportunity for losing luggage within the destination airport, your chance of ultimately receiving your bag should be identical to the probability of your bag being on the plane, i.e. on the flight segment to your final destination airport. More simply, if it is on the plane, then you will get it:

$$P(\textit{Your Bag on Carousel} = \textit{True} \mid \textit{Your Bag on Plane} = \textit{True}) = 1 \quad (4.2)$$

$$P(\textit{Your Bag on Carousel} = \textit{False} \mid \textit{Your Bag on Plane} = \textit{True}) = 0 \quad (4.3)$$

Conversely, the following must hold too:

$$P(\textit{Your Bag on Carousel} = \textit{False} \mid \textit{Your Bag on Plane} = \textit{False}) = 1 \quad (4.4)$$

$$P(\textit{Your Bag on Carousel} = \textit{True} \mid \textit{Your Bag on Plane} = \textit{False}) = 0 \quad (4.5)$$

We now encode this knowledge into our network. We add a second node, *Your Bag on Carousel* and then click the **Arc Creation Mode** icon ( ). Next, we click and hold the cursor on *Your Bag on Plane*, drag the cursor to *Your Bag on Carousel*, and finally release. This produces a simple, manually specified Bayesian network (Figure 4.3).

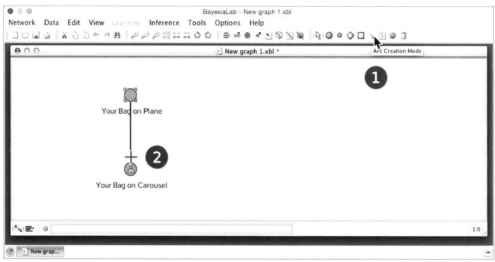

Figure 4.3

The yellow warning triangle ( ) indicates that probabilities need to be defined for the node *Your Bag on Carousel*. As opposed to the previous instance, where we only had to enter marginal probabilities, we now need to define the probabilities of the states of the node *Your Bag on Carousel* conditional on the states of *Your Bag on Plane*. In other words, we need to fill the **Conditional Probability Table** to quantify this parent-child relationship. We open the **Node Editor** and enter the values from the equations above.

Chapter 4

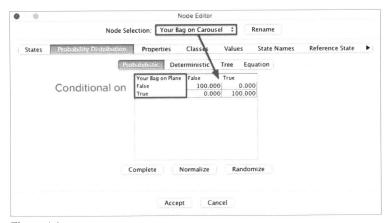

Figure 4.4

Introduction of Time

Now we add another piece of contextual information, which has not been mentioned yet in our story. From the baggage handler who monitors the carousel, you learn that 100 pieces of luggage in total were on your final flight segment, from the hub to the destination. After you wait for one minute, 10 bags have appeared on the carousel, and they keep coming out at a very steady rate. However, yours is not among the first ten that were delivered in the first minute. At the current rate, it would now take 9 more minutes for all bags to be delivered to the baggage carousel.

Given that your bag was not delivered in the first minute, what is your new expectation of ultimately getting your bag? How about after the second minute of waiting? Quite obviously, we need to introduce a time variable into our network. We create a new node *Time* and define discrete time intervals [0,...,10] to serve as its states.

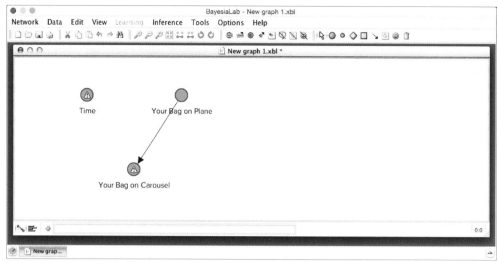

Figure 4.5

By default, all new nodes initially have two states, *True* and *False*. We can see this by opening the **Node Editor** and selecting the **States** tab (Figure 4.6).

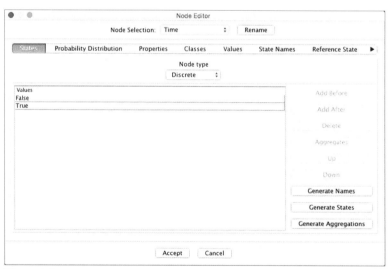

Figure 4.6

By clicking on the **Generate States** button, we create the states we need for our purposes. Here, we define 11 states, starting at 0 and increasing by 1 step (Figure 4.7).

Chapter 4

Figure 4.7

The **Node Editor** now shows the newly-generated states (Figure 4.8).

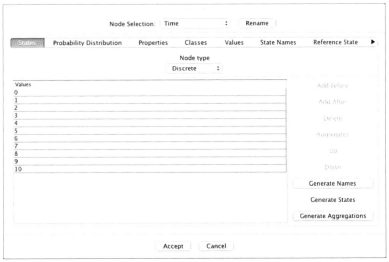

Figure 4.8

Beyond defining the states of *Time*, we also need to define their marginal probability distribution. For this, we select the tab **Probability Distribution > Probabilistic**. Quite naturally, no time interval is more probable than another one, so we should apply a uniform distribution across all states of *Time*. BayesiaLab provides a convenient shortcut for this purpose. Clicking the **Normalize** button places a uniform distribution across all cells, i.e. 9.091% per cell.

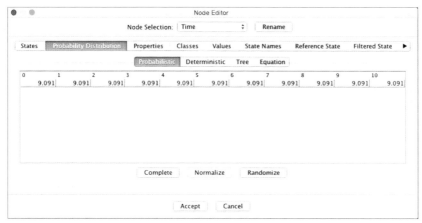

Figure 4.9

Once *Time* is defined, we draw an arc from *Time* to *Your Bag on Carousel*. By doing so, we introduce a causal relationship, stating that *Time* has an influence on the status of your bag.

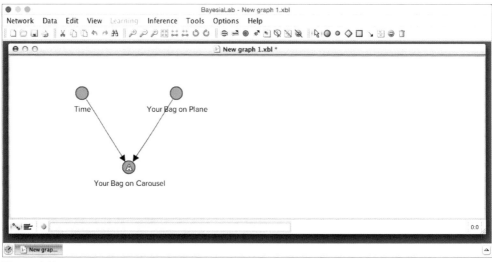

Figure 4.10

The warning triangle (⚠) once again indicates that we need to define further probabilities concerning *Your Bag on Carousel*. We open the **Node Editor** to enter these probabilities into the **Conditional Probability Table** (Figure 4.11).

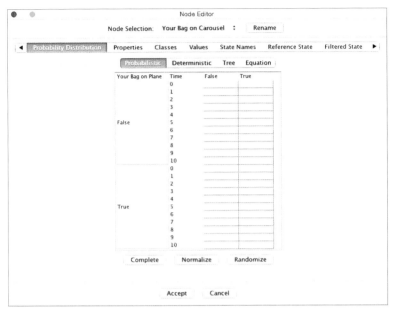

Figure 4.11

Note that the probabilities of the states *True* and *False* now depend on two parent nodes. For the upper half of the table, it is still quite simple to establish the probabilities. If the bag is not on the plane, it will not appear on the baggage carousel under any circumstance, regardless of *Time*. Hence, we set *False* to 100 (%) for all rows in which *Your Bag on Plane=False* (Figure 4.12).

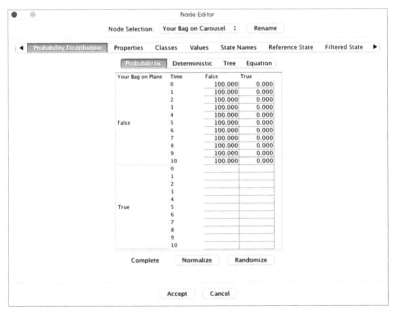

Figure 4.12

However, given that *Your Bag on Plane=True*, the probability of seeing it on the carousel depends on the time elapsed. Now, what is the probability of seeing your bag at each time step? Assuming that all luggage is shuffled extensively through the loading and unloading processes, there is a uniform probability distribution that the bag is anywhere in the pile of luggage to be delivered to the carousel. As a result, there is a 10% chance that your bag is delivered in the first minute, i.e. within the first batch of 10 out of 100 luggage pieces. Over the period of two minutes, there is a 20% probability that the bag arrives and so on. Only when the last batch of 10 bags remains undelivered, we can be certain that your bag is in the final batch, i.e. there is a 100% probability of the state *True* in the tenth minute. We can now fill out the **Conditional Probability Table** in the **Node Editor** with these values. Note that we only need to enter the values in the *True* column and then highlight the remaining empty cells. Clicking **Complete** prompts BayesiaLab to automatically fill in the *False* column to achieve a row sum of 100% (Figure 4.13).

Figure 4.13

Now we have a fully specified Bayesian network, which we can evaluate immediately.

Evidential Reasoning for Problem #1

BayesiaLab's **Validation Mode** provides the tools for using the Bayesian network we built for omni-directional inference. We switch to the **Validation Mode** via the cor-

Chapter 4

responding icon (≢), in the lower left-hand corner of the main window, or via the keyboard shortcut [F5] (Figure 4.14).

Figure 4.14

Upon switching to this mode, we double-click on all three nodes to bring up their associated **Monitors**, which show the nodes' current marginal probability distributions. We find these **Monitors** inside the **Monitor Panel** on the right-hand side of the main window (Figure 4.15).

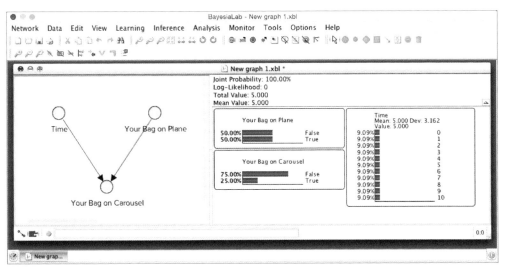

Figure 4.15

Inference Tasks

If we filled the **Conditional Probability Table** correctly, we should now be able to validate at least the trivial cases straight away, e.g. for *Your Bag on Plane=False*.

Inference from Cause to Effect: Your Bag on Plane=False

We perform inference by setting such evidence via the corresponding **Monitor** in the **Monitor Panel**. We double-click the bar that represents the **State** *False* (Figure 4.16).

Figure 4.16

The setting of the evidence turns the node and the corresponding bar in the **Monitor** green (Figure 4.17).

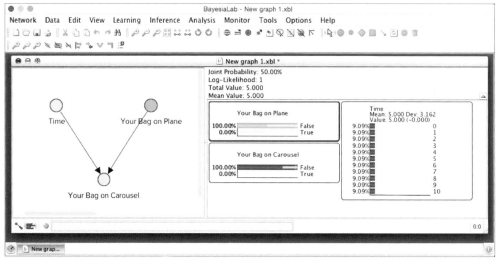

Figure 4.17

The **Monitor** for *Your Bag on Carousel* shows the result. The small gray arrows overlaid on top of the horizontal bars furthermore indicate how the probabilities have changed by setting this most recent piece of evidence (Figure 4.18).

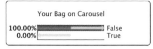

Figure 4.18

Indeed, your bag could not possibly be on the carousel because it was not on the plane in the first place. The inference we performed here is indeed trivial, but it is reassuring to see that the Bayesian network properly "plays back" the knowledge we entered earlier.

Omni-Directional Inference: Your Bag on Carousel=False, Time=1

The next question, however, typically goes beyond our intuitive reasoning capabilities. We wish to infer the probability that your bag made it onto the plane, given that we are now in minute 1, and the bag has not yet appeared on the carousel. This inference is tricky because we now have to reason along multiple paths in our network.

Diagnostic Reasoning

The first path is from *Your Bag on Carousel* to *Your Bag on Plane*. This type of reasoning from effect to cause is more commonly known as diagnosis. More formally, we can write:

$$P(Your\ Bag\ on\ Plane = True\ |\ Your\ Bag\ on\ Carousel = False) \tag{4.6}$$

Inter-Causal Reasoning

The second reasoning path is from *Time* via *Your Bag on Carousel* to *Your Bag on Plane*. Once we condition on *Your Bag on Carousel*, i.e. by observing the value, we open this path, and information can flow from one cause, *Time*, via the common effect,[1] *Your Bag on Carousel*, to the other cause, *Your Bag on Plane*. Hence, we speak of "inter-causal reasoning" in this context. The specific computation task is:

▸ Common Child (Collider) in Chapter 10, p. 337.

$$P(Your\ Bag\ on\ Plane = True\ |\ Your\ Bag\ on\ Carousel = False, Time = 1) \tag{4.7}$$

Bayesian Networks as Inference Engine

How do we go about computing this probability? We do not attempt to perform this computation ourselves. Rather, we rely on the Bayesian network we built and BayesiaLab's exact inference algorithms. However, before we can perform this inference computation, we need to remove the previous piece of evidence, i.e. *Your Bag on Plane=True*. We do this by right-clicking the relevant node and then selecting **Remove Evidence** from the **Contextual Menu** (Figure 4.19). Alternatively, we can remove all evidence by clicking the **Remove All Observations** icon ( ).

---

1 Chapter 10 will formally explain the specific causal roles nodes can play in a network, such as "common effect," along with their implications for observational and causal inference.

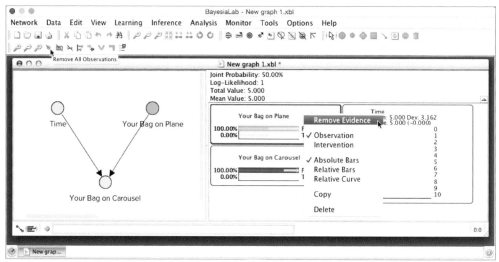

Figure 4.19

Then, we set the new observations via the **Monitors** in the **Monitor Panel**. The inference computation then happens automatically (Figure 4.20).

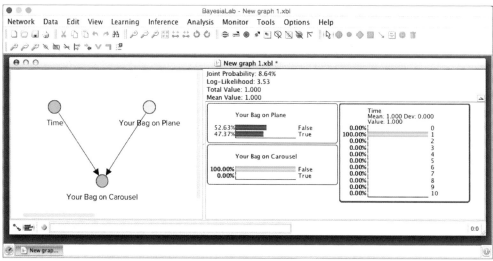

Figure 4.20

Given that you do not see your bag in the first minute, the probability that your bag made it onto the plane is now no longer at the marginal level of 50%, but is reduced to 47.37%.

Inference as a Function of Time

Continuing with this example, how about if the bag has not shown up in the second minute, in the third minute, etc.? We can use one of BayesiaLab's built-in visualization functions to analyze this automatically. To prepare the network for this type of analysis, we first need to set a **Target Node**, which, in our case, is *Your Bag on Plane*. Upon right-clicking this node, we select **Set as Target Node** (Figure 4.21). Alternatively, we can double-click the node, or one of its states in the corresponding **Monitor**, while holding ⊤.

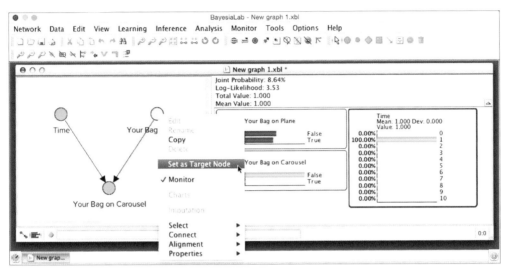

Figure 4.21

Upon setting the **Target Node**, *Your Bag on Plane* is marked with a bullseye symbol (◎). Also, the corresponding **Monitor** is now highlighted in red (Figure 4.22).

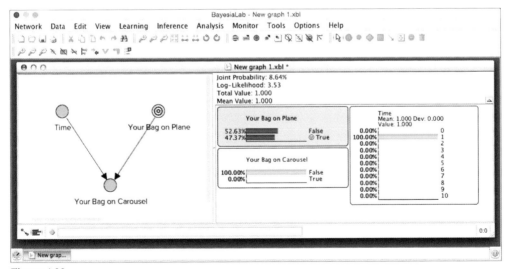

Figure 4.22

Before we continue, however, we need to remove the evidence from the *Time* **Monitor**. We do so by right-clicking the **Monitor** and selecting **Remove Evidence** from the **Contextual Menu** (Figure 4.23).

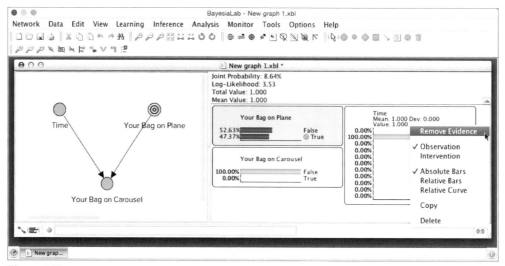

Figure 4.23

Then, we select **Analysis > Visual > Influence Analysis on Target Node** (Figure 4.24).

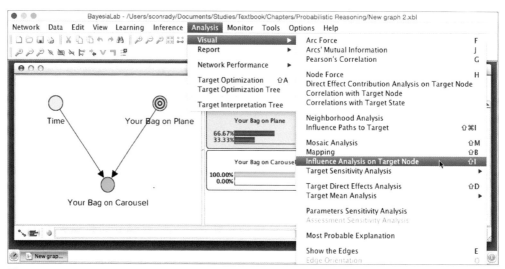

Figure 4.24

The resulting graph shows the probabilities of receiving your bag as a function of the discrete time steps. To see the progression of the *True* state, we select the corresponding tab at the top of the window (Figure 4.25).

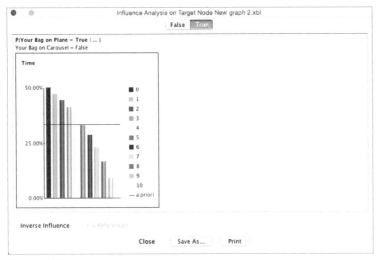

Figure 4.25

Knowledge Modeling for Problem #2

Continuing with our narrative, you now notice a colleague of yours in the baggage claim area. As it turns out, your colleague was traveling on the very same itinerary as you, so he had to make the same tight connection. As opposed to you, he has already retrieved his bag from the carousel. You assume that his luggage being on the airplane is not independent of your luggage being on the same plane, so you take this as a positive sign. How do we formally integrate this assumption into our existing network?

To encode any new knowledge, we first need to switch back into the **Modeling Mode** (or F4). Then, we duplicate the existing nodes *Your Bag on Plane* and *Your Bag on Carousel* by copying and pasting them using the common shortcuts, Control C and Control V, into the same graph (Figure 4.26).

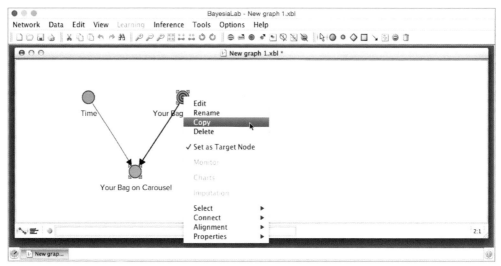

Figure 4.26

In the copy process, BayesiaLab prompts us for a **Copy Format** (Figure 4.26), which would only be relevant if we intended to paste the selected portion of the network into another application, such as PowerPoint. As we paste the copied nodes into the same **Graph Panel**, the format does not matter.

Figure 4.27

Upon pasting, by default, the new nodes have the same names as the original ones plus the suffix "[1]" (Figure 4.28).

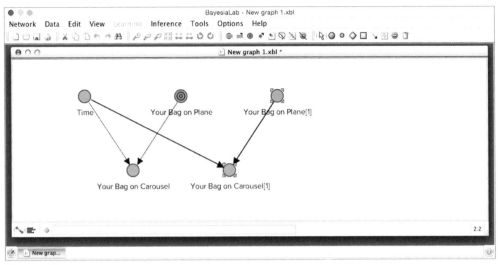

Figure 4.28

Next, we reposition the nodes on the **Graph Panel** and rename them to show that the new nodes relate to your colleague's situation, rather than yours. To rename the nodes we double-click the **Node Names** and overwrite the existing label.

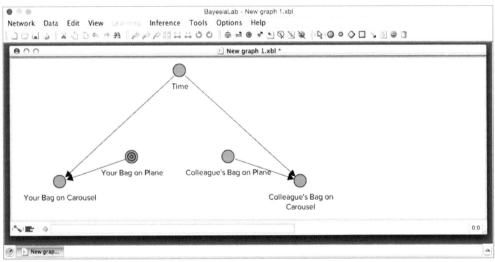

Figure 4.29

The next assumption is that your colleague's bag is subject to exactly the same forces as your luggage. More specifically, the successful transfer of your and his luggage is a function of how many bags could be processed at the hub airport given the limited transfer time. To model this, we introduce a new node and name it *Transit* (Figure 4.30).

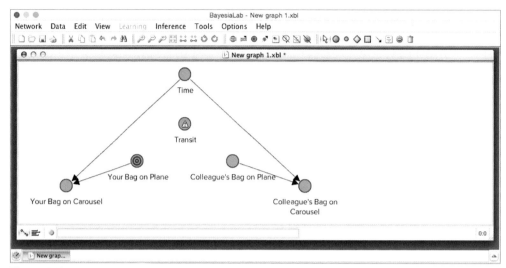

Figure 4.30

We create 7 states of ten-minute intervals for this node, which reflect the amount of time available for the transfer, i.e. from 0 to 60 minutes (Figure 4.31).

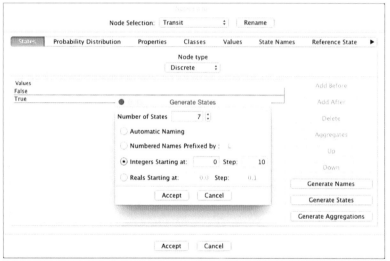

Figure 4.31

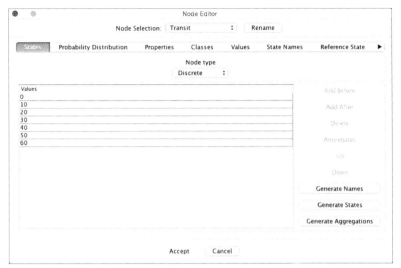

Figure 4.32

Furthermore, we set the probability distribution for *Transit*. For expository simplicity, we apply a uniform distribution using the **Normalize** button (Figure 4.33).

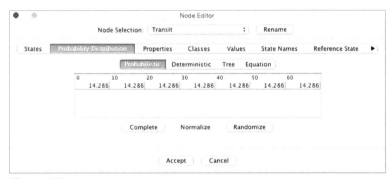

Figure 4.33

Now that the *Transit* node is defined, we can draw the arcs connecting it to *Your Bag on Plane* and *Colleague's Bag on Plane* (Figure 4.34).

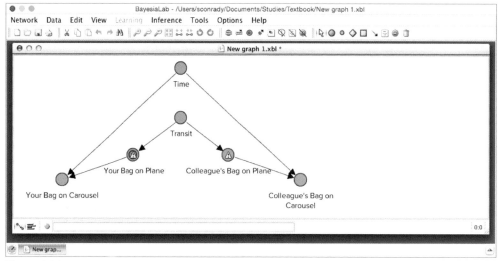

Figure 4.34

The yellow warning triangles (⚠) indicate that the conditional probability tables of *Your Bag on Plane* and *Colleague's Bag on Plane* have yet to be filled. Thus, we need to open the **Node Editor** and set these probabilities. We will assume that the probability of your bag making the connection is 0% given a *Transit* time of 0 minutes and 100% with a *Transit* time of 60 minutes. Between those values, the probability of a successful transfer increases linearly with time (Figure 4.35).

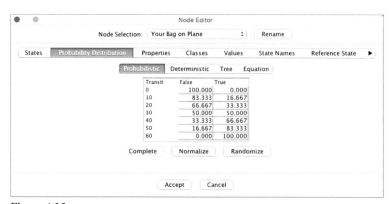

Figure 4.35

The very same function also applies to your colleague's bag, so we enter the same conditional probabilities for the node *Colleague's Bag on Plane* by copying and pasting the previously entered table (Figure 4.35). Figure 4.36 shows the completed network.

Chapter 4

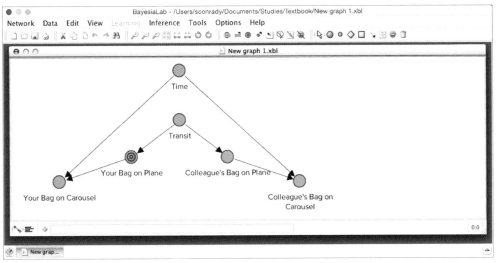

Figure 4.36

Evidential Reasoning for Problem #2

Now that the probabilities are defined, we switch to the **Validation Mode** (☰ or [F5]); our updated Bayesian network is ready for inference again (Figure 4.37).

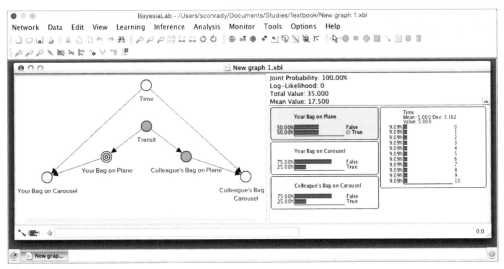

Figure 4.37

We simulate a new scenario to test this new network. For instance, we move to the fifth minute and set evidence that your bag has not yet arrived (Figure 4.38).

73

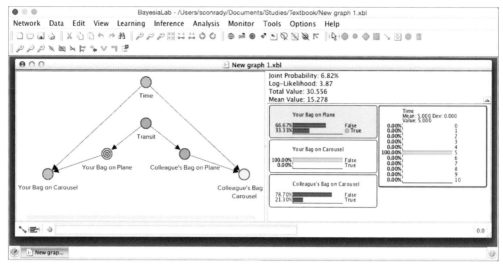

Figure 4.38

Given these observations, the probability of *Your Bag on Plane=True* is now 33.33%. Interestingly, the probability of *Colleague's Bag on Plane* has also changed. As evidence propagates omni-directionally through the network, our two observed nodes do indeed influence *Colleague's Bag on Plane*. A further iteration of the scenario in our story is that we observe *Colleague's Bag on Carousel=True*, also in the fifth minute (Figure 4.39).

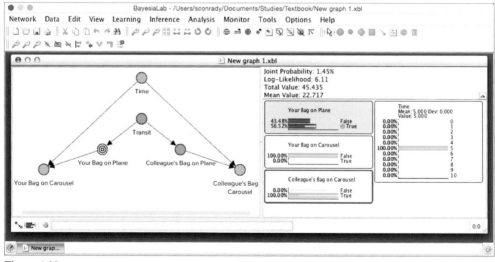

Figure 4.39

Given the observation of *Colleague's Bag on Carousel*, even though we have not yet seen *Your Bag on Carousel*, the probability of *Your Bag on Plane* increases to 56.52%. Indeed, this observation should change your expectation quite a bit. The small gray arrows on the blue bars inside the **Monitor** for *Your Bag on Plane* indicate the impact of this observation.

After removing the evidence from the *Time* **Monitor**, we can perform **Influence Analysis on Target** again in order to see the probability of *Your Bag on Plane=True* as a function of *Time*, given *Your Bag on Carousel=False* and *Colleague's Bag on Carousel=True* (Figure 4.41). To focus our analysis on *Time* alone, we select the *Time* node and then select **Analysis > Visual > Influence Analysis on Target** (Figure 4.40).

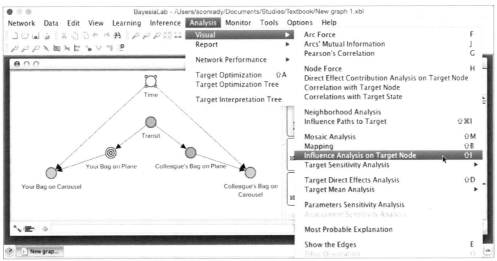

Figure 4.40

As before, we select the *True* tab in the resulting window to see the evolution of probabilities given *Time* (Figure 4.41).

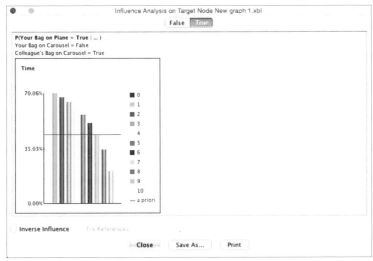

Figure 4.41

Summary

This chapter provided a brief introduction to knowledge modeling and evidential reasoning with Bayesian networks in BayesiaLab. Bayesian networks can formally encode available knowledge, deal with uncertainties, and perform omni-directional inference. As a result, we can properly reason about a problem domain despite many unknowns.

Chapter 4

Chapter 5

5. Bayesian Networks and Data

In the previous chapter, we described the application of Bayesian networks for evidential reasoning. In that example, all available knowledge was manually encoded in the Bayesian network. In this chapter, we additionally use data for defining Bayesian networks. This provide the basis for the following chapters, which will present applications that utilize machine-learning for generating Bayesian networks entirely from data. For machine learning with BayesiaLab, concepts derived from information theory, such as entropy and mutual information, are of particular importance and should be understood by the researcher. However, to most scientists these measures are not nearly as familiar as common statistical measures, e.g. covariance and correlation.

Example: House Prices in Ames, Iowa

To introduce these presumably unfamiliar information-theoretic concepts, we present a straightforward research task. The objective is to establish the predictive importance of a range of variables with regard to a target variable. The domain of this example is residential real estate, and we wish to examine the relationships between home characteristics and sales price. In this context, it is natural to ask questions related to variable importance, such as, which is the most important predictive variable pertaining to home value? By attempting to answer this question, we can explain what entropy and mutual information mean in practice and how BayesiaLab computes these measures. In this process, we also demonstrate a number of BayesiaLab's data handling functions.

The dataset for this chapter's exercise describes the sale of individual residential properties in Ames, Iowa, from 2006 to 2010. It contains a total of 2,930 observations and a large number of explanatory variables (23 nominal, 23 ordinal, 14 discrete, and 20 continuous). This dataset was first used by De Cock (2011) as an educational tool for statistics students. The objective of their study was the same as ours, i.e. modeling sale prices as a function of the property attributes.

To make this dataset more convenient for demonstration purposes, we reduced the total number of variables to 49. This pre-selection was fairly straightforward as there are numerous variables that essentially do not apply to homes in Ames, e.g. variables relating to pool quality and pool size (there are practically no pools), or roof material (it is the same for virtually all homes).

Data Import Wizard

As the first step, we start BayesiaLab's **Data Import Wizard** by selecting **Data > Open Data Source > Text File** from the main menu.[1]

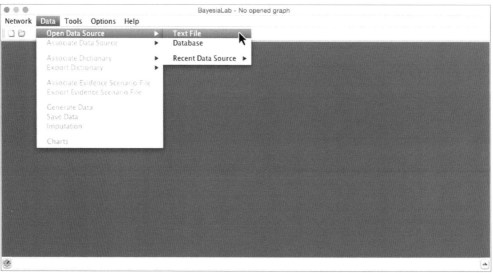

Figure 5.1

Next, we select the file named "ames.csv", a comma-delimited, flat text file.[2]

---

1 For larger datasets, we could use **Data > Open Data Source > Database** and connect to a database server via BayesiaLab's JDBC connection.

2 The Ames dataset is available for download via this link: www.bayesia.us/ames

Chapter 5

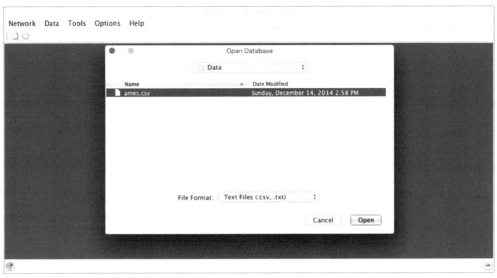

Figure 5.2

This brings up the first screen of the **Data Import Wizard**, which provides a preview of the to-be-imported dataset (Figure 5.3). For this example, the coding options for **Missing Values** and **Filtered Values** are particularly important. By default, Bayesia-Lab lists commonly used codes that indicate an absence of data, e.g. #NUL! or NR (non-response). In the Ames dataset, a blank field ("") indicates a **Missing Value**, and "FV" stands for **Filtered Value**. These are recognized automatically. If other codes were used, we could add them to the respective lists on this screen.

▶ Chapter 9. Missing Values Processing, p. 289.

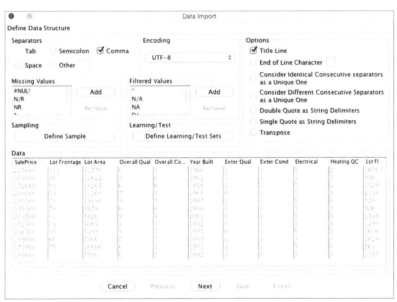

Figure 5.3

Clicking **Next**, we proceed to the screen that allows us to define variable types (Figure 5.4).

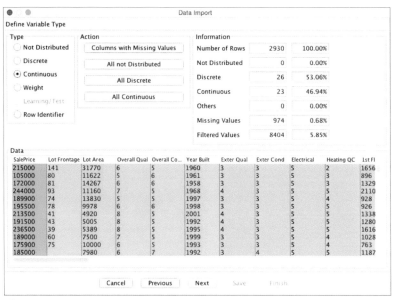

Figure 5.4

BayesiaLab scans all variables in the database and comes up with a best guess regarding the variable type (Figure 5.4). Variables identified as **Continuous** are shown in turquoise, and those identified as **Discrete** are highlighted in pastel red.

In BayesiaLab, a **Continuous** variable contains a wide range of numerical values (discrete or continuous), which need to be transformed into a more limited number of discrete states. Some other variables in the database only have very few distinct numerical values to begin with, e.g. [1,2,3,4,5], and BayesiaLab automatically recognizes such variables as **Discrete**. For them, the number of numerical states is small enough that it is not necessary to create bins of values. Also, variables containing text values are automatically considered **Discrete**.

For this dataset, however, we need to make a number of adjustments to BayesiaLab's suggested data types. For instance, we set all numerical variables to **Continuous**, including those highlighted in red that were originally identified as **Discrete**. As a result, all columns in the data preview of the **Data Import Wizard** are now shown in turquoise (Figure 5.5).

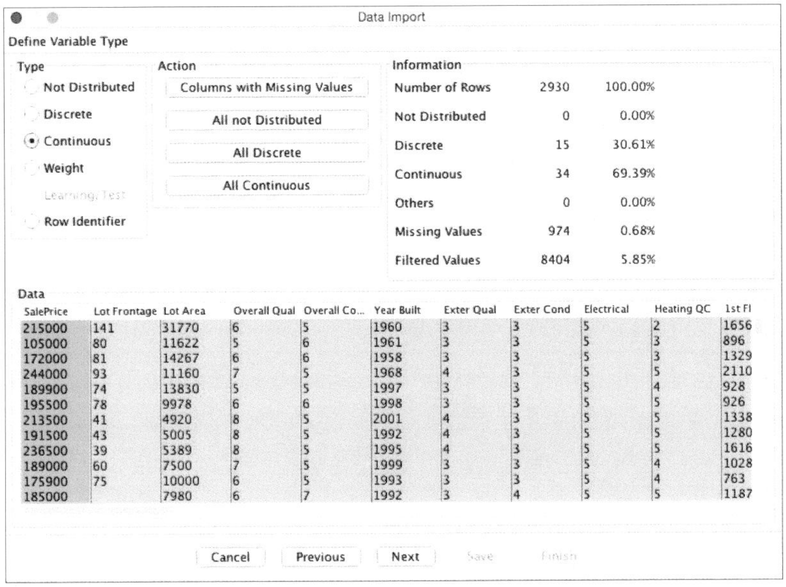

Figure 5.5

Given that our database contains some missing values, we need to select the type of **Missing Values Processing** in the next step (Figure 5.6). Instead of using ad hoc methods, such as pairwise or listwise deletion, BayesiaLab can leverage more sophisticated techniques and provide estimates (or temporary placeholders) for such missing values—without discarding any of the original data. We will discuss **Missing Values Processing** in detail in Chapter 9. For this example, however, we leave the default setting of **Structural EM**.

▶ Chapter 9. Missing Values Processing, p. 289.

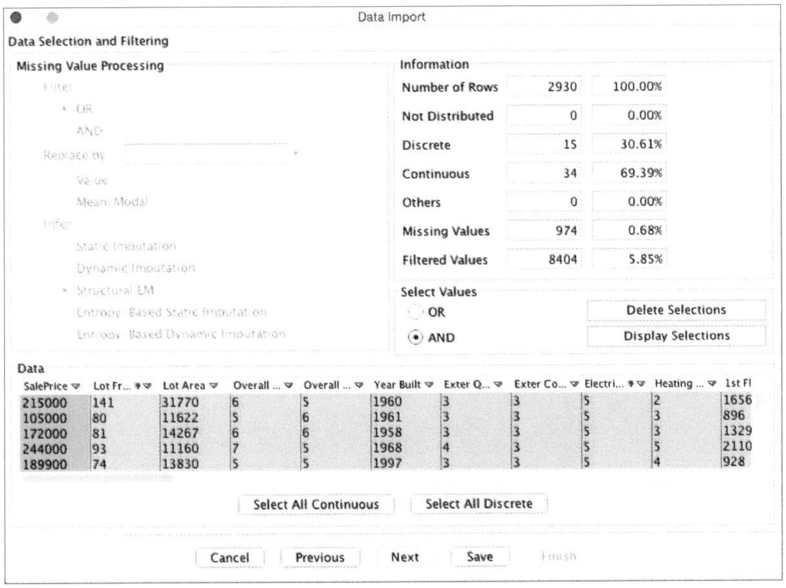

Figure 5.6

Filtered Values

> Filtered Values in Chapter 9, p. 296.

At this juncture, however, we need to introduce a very special type of missing value for which we *must not* generate any estimates. We are referring to so-called **Filtered Values**. These are "impossible" values that *do not or cannot exist*—given a specific set of evidence, as opposed to values that *do exist* but are not observed. For example, for a home that does not have a garage, there cannot be any value for the variable *Garage Type*, such as *Attached to Home, Detached from Home,* or *Basement Garage*. Quite simply, if there is no garage, there cannot be a garage type. As a result, it makes no sense to calculate an estimate of a **Filtered Value**. In a database, unfortunately, a **Filtered Value** typically looks identical to "true" missing value that *does exist but is not observed*. The database typically contains the same code, such as a blank, NULL, N/A, etc., for both cases.

Therefore, as opposed to "normal" missing values, which can be left as-is in the database, we must mark **Filtered Values** with a specific code, e.g. "*FV.*" The **Filtered Value** declaration should be done during data preparation, prior to importing any data into BayesiaLab. BayesiaLab will then add a **Filtered State** (marked with "*") to the discrete states of the variables with **Filtered Values**, and utilize a special approach for actively disregarding such **Filtered States**, so that they are *not* taken into account during machine-learning or for estimating effects.

Discretization

As the next step in the **Data Import Wizard**, all **Continuous** values must be discretized (or binned). We show a sequence of screenshots to highlight the necessary steps. The initial view of the **Discretization and Aggregation** step appears in Figure 5.7.

Chapter 5

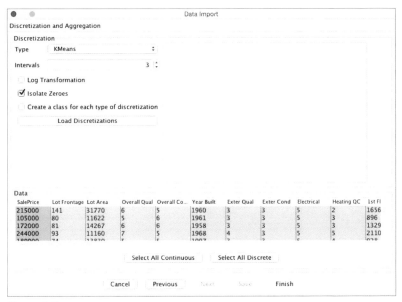

Figure 5.7

By default, the first column is highlighted, which happens to be *SalePrice*, the variable of principal interest in this example. Instead of selecting any of the available automatic discretization algorithms, we pick **Manual** from the **Type** drop-down menu, which brings up the cumulative distribution function of the *SalePrice* variable (Figure 5.8).

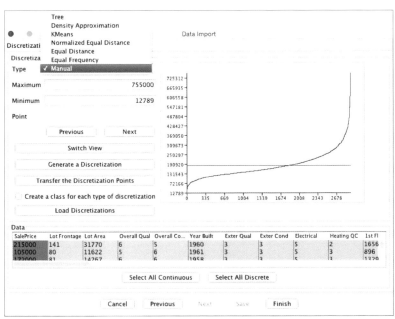

Figure 5.8

By clicking **Switch View**, we can bring up the probability density function of *Sale-Price* (Figure 5.9).

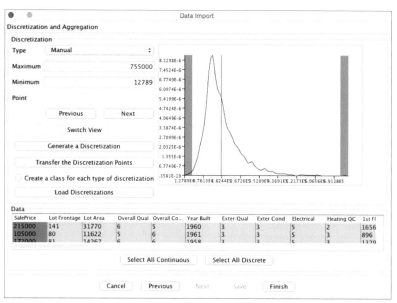

Figure 5.9

Either view allows us to examine the distribution and identify any salient points. We stay on the current screen to set the thresholds for each discretization bin (Figure 5.10). In many instances, we would use an algorithm to define bins automatically, unless the variable will serve as the target variable. In that case, we usually rely on available expert knowledge to define the binning. In this example, we wish to have evenly-spaced, round numbers for the interval boundaries. We add boundaries by right-clicking on the plot (right-clicking on an existing boundary removes it again). Furthermore, we can fine-tune a threshold's position by entering a precise value in the **Point** field. We use {75000, 150000, 225000, 300000} as the interval boundaries (Figure 5.10).

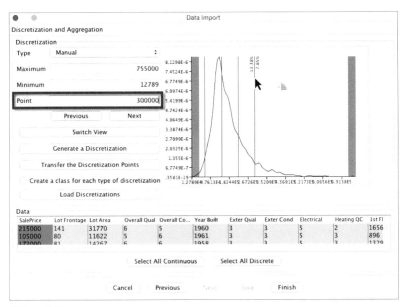

Figure 5.10

Now that we have manually discretized the target variable *SalePrice* (column highlighted in blue in Figure 5.10), we still need to discretize the remaining continuous variables. However, we will take advantage of an automatic discretization algorithm for those variables. We click **Select All Continuous** and then deselect *SalePrice* by clicking on the corresponding column while holding [Control]. This excludes *SalePrice* from the subsequent automatic discretization.

Different discretization algorithms are available, five univariate that only use the data of the to-be-discretized variable, and one bivariate that uses the data of the to-be-discretized variable plus the data of a target variable:

Equal Distance

Equal Distance uses the range of the variable to define an equal repartition of the discretization thresholds. This method is particularly useful for discretizing variables that share the same variation domain (e.g. satisfaction measures in surveys). Additionally, this method is suitable for obtaining a discrete representation of the density function. However, it is extremely sensitive to outliers, and it can return bins that do not contain any data points.

Normalized Equal Distance

Normalized Equal Distance pre-processes the data with a smoothing algorithm to remove outliers prior to defining equal partitions.

Equal Frequency

With **Equal Frequency**, the discretization thresholds are computed with the objective of obtaining bins with the same number of observations, which usually results in a uniform distribution. Thus, the shape of the original density function is no longer apparent upon discretization. As we will see later in this chapter, this also leads to an artificial increase in the entropy of the system, which has a direct impact on the complexity of machine-learned models. However, this type of discretization can be useful—once a structure is learned—for further increasing the precision of the representation of continuous values.

K-Means

K-Means is based on the classical K-Means data clustering algorithm but uses only one dimension, which is to-be-discretized variable. **K-Means** returns a discretization that is directly dependent on the density function of the variable. For example, applying a three-bin **K-Means** discretization to a normally distributed variable creates a central bin representing 50% of the data points, along with two bins of 25% each for the tails. In the absence of a target variable, or if little else is known about the variation domain and distribution of the continuous variables, **K-Means** is recommended as the default method.

Tree

Tree is a bivariate discretization method. It machine learns a tree that uses the to-be-discretized variable for representing the conditional probability distributions of the target variable given the to-be-discretized variable. Once the tree learned, it is analyzed to extract the most useful thresholds. This is the method of choice in the context of **Supervised Learning**, i.e. when planning to machine-learn a model to predict the target variable.

At the same time, we do not recommend using **Tree** in the context of **Unsupervised Learning**. The **Tree** algorithm creates bins that have a bias towards the

Chapter 5

designated target variable. Naturally, emphasizing one particular variable would run counter the intent of **Unsupervised Learning**.

Note that if the to-be-discretized variable is independent of the target variable, it will be impossible to build a tree and BayesiaLab will prompt for the selection of a univariate discretization algorithm.

In this example, we focus our analysis on *SalePrice*, which can be considered a type of **Supervised Learning**. Therefore, we choose to discretize all continuous variables with the **Tree** algorithm, using *SalePrice* as the **Target** variable. The **Target** must either be a **Discrete** variable or a **Continuous** variable that has already been manually discretized, which is the case for *SalePrice*.

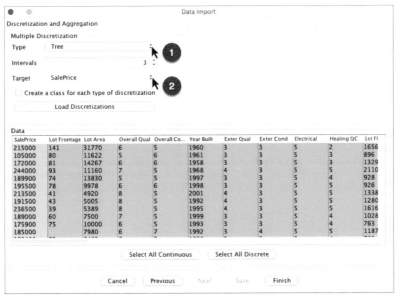

Figure 5.11

Once this is set, clicking **Finish** completes the import process. The import process concludes with a pop-up window that offers to display the **Import Report** (Figure 5.12).

Figure 5.12

Clicking **Yes** brings up the **Import Report**, which can be saved in HTML format. It lists the discretization intervals of **Continuous** variables, the **States** of **Discrete** variables, and the discretization method that was used for each variable (Figure 5.13).

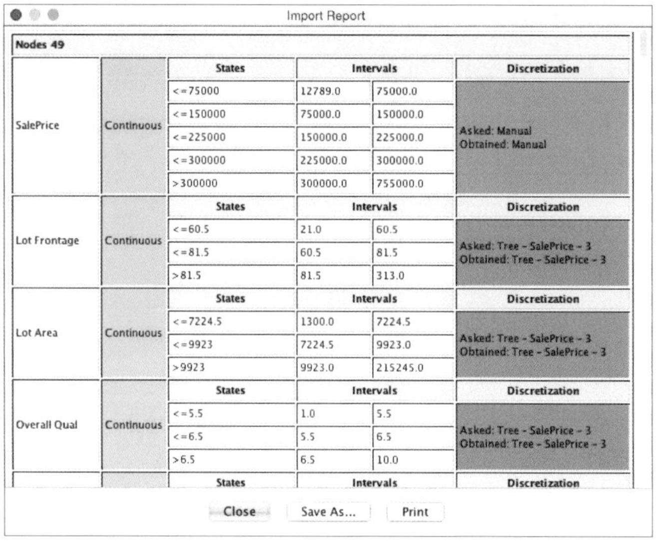

Figure 5.13

Graph Panel

Once we close out of this report, we can see the result of the import. Nodes in the **Graph Panel** now represent all the variables from the imported database (Figure 5.14). The dashed borders of some nodes (◯) indicate that the corresponding variables were discretized during data import. Furthermore, we can see icons that indicate the presence of **Missing Values** (?) and **Filtered Values** () on the respective nodes.

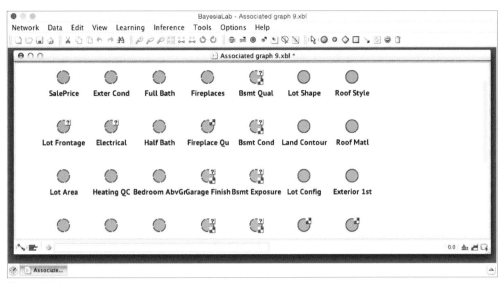

Figure 5.14

The lack of warning icons on any of the nodes indicates that all their parameters, i.e. their marginal probability distributions, were automatically estimated upon data import. To verify, we can double-click *SalePrice*, go to the **Probability Distribution | Probabilistic** tab, and see this node's marginal distribution.

Figure 5.15

Clicking on **Occurrences** tab shows the observations per cell, which were used for the **Maximum Likelihood Estimation** of the marginal distribution.

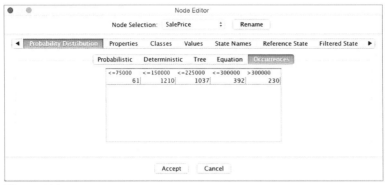

Figure 5.16

Node Comments

The **Node Names** that are displayed by default were taken directly from the column header in the dataset. Given their typical brevity, to keep the **Graph Panel** uncluttered, we like to keep the column headers from the database as **Node Names**. On the other hand, we may wish to have longer, more descriptive names available as **Node Comments** when interpreting the network. These comments can be edited via the **Node Editor**. Alternatively, we can create a **Dictionary**, i.e. a file that links **Node Names** to **Node Comments**.

The syntax for this association is rather straightforward: we simply define a text file that includes one **Node Name** per line. Each **Node Name** is followed by a delimiter ("=", tab, or space) and then by the long node description, which will serve as **Node Comment** (Figure 5.17). Note that basic HTML tags can be included in the dictionary file.

```
Node Comments 3.txt
SalePrice=Sale Price
Lot Frontage=Lot Frontage (Continuous): Linear feet of street connected to property
Lot Area=Lot Area (Continuous): Lot size in square feet
Overall Qual=Overall Qual (Ordinal): Rates the overall material and finish of the house
Overall Cond=Overall Cond (Ordinal): Rates the overall condition of the house
Year Built=Year Built (Discrete): Original construction date
Exter Qual=Exter Qual (Ordinal): Evaluates the qualityof the material on the exterior
Exter Cond=Exter Cond (Ordinal): Evaluates the present condition of the material on the exterior
Electrical=Electrical (Ordinal): Electrical system
Heating QC=HeatingQC (Ordinal): Heating quality and condition
1st Flr SF=1st Flr SF (Continuous): First Floor square feet
2nd Flr SF=2nd Flr SF (Continuous): Second floor square feet
Gr Liv Area=Gr Liv Area (Continuous): Above grade (ground) living area square feet
Bsmt Full Bath=Bsmt Full Bath (Discrete): Basement full bathrooms
Full Bath=Full Bath (Discrete): Full bathrooms above grade
Half Bath=Half Bath (Discrete): Half baths above grade
Bedroom AbvGr=Bedroom (Discrete): Bedrooms above grade (does NOT include basement bedrooms)
Kitchen AbvGr=Kitchen (Discrete): Kitchens above grade
Kitchen Qual=KitchenQual (Ordinal): Kitchen quality
TotRms AbvGrd=TotRmsAbvGrd (Discrete): Total rooms above grade (does not include bathrooms)
Functional=Functional (Ordinal): Home functionality (Assume typical unless deductions are warranted)
Fireplaces=Fireplaces (Discrete): Number of fireplaces
Fireplace Qu=FireplaceQu (Ordinal): Fireplace quality
Garage Finish=arage Finish (Ordinal): Interior finish of the garage
Garage Cars=Garage Cars (Discrete): Size of garage in car capacity
Garage Area=Garage Area (Continuous): Size of garage in square feet
```

Figure 5.17

To attach this **Dictionary**, we select **Data > Associate Dictionary > Node > Comments** (Figure 5.18).

Chapter 5

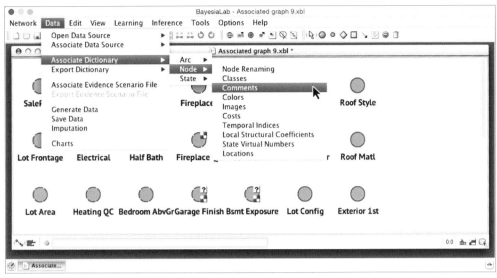

Figure 5.18

Next, we select the location of the **Dictionary** file, which is appropriately named "Node Comments.txt" (Figure 5.19).

Figure 5.19

Upon loading the **Dictionary** file, a call-out icon (◉) appears next to each node. This means that a **Node Comment** is available. They can be displayed by clicking the **Display Node Comment** icon (◉) in the menu bar (Figure 5.20). **Node Comments** can be turned on or off, depending on the desired view.

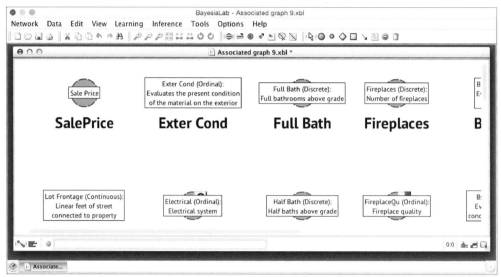

Figure 5.20

We now switch to the **Validation Mode** (or), in which we can bring up individual **Monitors** by double-clicking the nodes of interest. We can also select multiple nodes and then double-click any one of them to bring up all of their **Monitors**. (Figure 5.21).

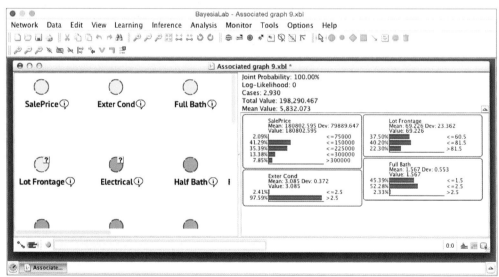

Figure 5.21

Now that we have our database internally represented in BayesiaLab, we need to become familiar how BayesiaLab can quantify the probabilistic properties of these nodes and their relationships.

Chapter 5

Information-Theoretic Concepts

Uncertainty, Entropy, and Mutual Information

In a traditional statistical analysis, we would presumably examine correlation and covariance between the variables to establish their relative importance, especially with regard to the target variable *Sale Price*. In this chapter, we take an alternative approach, which is based on information theory. Instead of computing the correlation coefficient, we consider how the uncertainty of the states of a to-be-predicted variable is affected by observing a predictor variable.

Beyond our common-sense understanding of uncertainty, there is a more formal quantification of uncertainty in information theory, and that is entropy. More specifically, we use entropy to quantify the uncertainty manifested in the probability distribution of a variable or of a set of variables. In the context of our example, the uncertainty relates to the to-be-predicted home price.

It is fair to say that we would need detailed information about a property to make a reasonable prediction of its value. However, in the absence of any specific information, would we be entirely uncertain about its value? Probably not. Even if we did not know anything about a particular house, we would have some contextual knowledge, i.e. that the house is in Ames, Iowa, rather than in midtown-Manhattan, and that the property is a private home rather than shopping mall. That knowledge significantly reduces the range of possible values. True uncertainty would mean that a value of $0.01 is as probable as a value of $1 million or $1 billion. That is clearly not the case here. So, how uncertain are we about the value of a random home in Ames, prior to learning anything about that particular home? The answer is that we can compute the entropy from the marginal probability distribution of home values in Ames. Since we have the Ames dataset already imported into BayesiaLab, we can display a histogram of *SalePrice* by bringing up its **Monitor** (Figure 5.22).

Figure 5.22

This **Monitor** reflects the discretization intervals that we defined during the data import. It is now easy to see the frequency of prices in each price interval, i.e. the marginal distribution of *SalePrice*. For instance, only about 2% of homes sold had a price of $75,000 or less. On the basis of this probability distribution, we can now compute the entropy. The definition of entropy for a discrete distribution is:

$$H(X) = -\sum_{x \in X} P(x) \log_2 P(x) \tag{5.1}$$

Entering the values displayed in the **Monitor**, we obtain:

$$H(SalePrice) = -(0.0208 \times \log_2(0.0208) + 0.413 \times \log_2(0.413) \tag{5.2}$$
$$+ 0.3539 \times \log_2(0.3539) + 0.1338 \times \log_2(0.1338)$$
$$+ 0.0785 \times \log_2(0.0785)) = 1.85$$

In information theory, the unit of information is bit, which is why we use base 2 of the logarithm. On its own, the calculated entropy value of 1.85 bits may not be a meaningful measure. To get a sense of how much or how little uncertainty this value represents, we compare it to two easily-interpretable measures, i.e. "no uncertainty" and "complete uncertainty."

No Uncertainty

No uncertainty means that the probability of one bin (or state) of *SalePrice* is 100%. This could be, for instance, *P(SalePrice<=75000)=1* (Figure 5.23).

Figure 5.23

We now compute the entropy of this distribution once again:

$$H(SalePrice_{<75,000}) = -(1 \times \log_2(1) + 0 \times \log_2(0) + 0 \times \log_2(0) + 0 \times \log_2(0)) \tag{5.3}$$
$$= \log_2(1) = 0$$

Here, $0 \times \log_2(0)$ is taken as 0, given the limit $\lim_{p \to 0+} p \log(p) = 0$. This simply means that "no uncertainty" has zero entropy.

Complete Uncertainty

What about the opposite end of the spectrum, i.e. complete uncertainty? Maximum uncertainty exists when all possible states of a distribution are equally probable, when we have a uniform distribution, as shown in Figure 5.24.

Figure 5.24

Once again, we calculate the entropy:

$$H(SalePrice_{uniform}) = -(0.2 \times \log_2(0.2) + 0.2 \times \log_2(0.2) \qquad (5.4)$$
$$+ 0.2 \times \log_2(0.2) + 0.2 \times \log_2(0.2) + 0.2 \times \log_2(0.2)) = \log_2(5) = 2.3219$$

In addition to the previously computed marginal entropy of 1.85, we now have the values 0 and 2.3219 for "no uncertainty" and "complete uncertainty" respectively.[3]

Entropy and Predictive Importance

How do such entropy values help us to establish the importance of predictive variables? If there is no uncertainty regarding a variable, one state of this variable has to have a 100% probability, and predicting that particular state must be correct. This would be like predicting the presence of clouds during rain. On the other hand, if the probability distribution of the target variable is uniform, e.g. the outcome of (fair) coin toss, a random prediction has to be correct with a probability of 50%.

In the context of house prices, knowing the marginal distribution of *SalePrice* and assuming this distribution is still true when we are making the prediction, predicting *SalePrice>=150000* would have a 41.28% probability of being correct, even if we knew nothing else. However, we would expect that observing an attribute of a specific home would reduce our uncertainty concerning its *SalePrice* and increase our probability of making a correct prediction for this particular home. In other words, conditional upon learning an attribute of a home, i.e. by observing a predictive variable, we expect a lower uncertainty for the target variable, *SalePrice*.

For instance, the moment we learn of a particular home that *LotArea=200,000* (measured in sq ft),[4] and assuming, again, that the estimated marginal distribution is still true when we are making the prediction, we can be certain that *SalePrice>300000*. This means that upon learning the value of this home's *LotArea*, the entropy of *SalePrice* goes from 1.85 to 0. Learning the size reduces our entropy by 1.85 bits. Alternatively, we can say that we gain information amounting to 1.85 bits.

---

3   The value 5 in the logarithm of the simplified equation (5.4) reflects the number of states. This means that the entropy is a function of the variable discretization.

4   200,000 square feet ≈ 4.6 acres ≈ 18.851 m^2 ≈1.86 ha.

The information gain or entropy reduction from learning about *LotArea* of this house is obvious. Observing a different home with a more common lot size, e.g. *LotArea=10,000,* would presumably provide less information and, thus, have less predictive value for that home.

Mutual Information

However, we wish to know how much information we would gain on average—considering all values of *LotArea* along with their probabilities—by generally observing it as a predictive variable for *SalePrice*. Knowing this "average information gain" would reflect the predictive importance of observing the variable *LotArea*.

To compute this, we need two quantities. First, the marginal entropy of the target variable *H(SalePrice)* (5.2), and, second, the conditional entropy of the target variable given the predictive variable:

$$H(SalePrice \mid LotArea) = \sum_i P(SalePrice_i) H(SalePrice_i \mid LotArea_i) \tag{5.5}$$

The difference between the marginal entropy of the target variable and the conditional entropy of the target given the predictive variable is formally known as **Mutual Information**, denoted by *I*. In our example, the **Mutual Information** *I* between *SalePrice* and *LotArea* is the marginal entropy of *SalePrice* minus the conditional entropy of *SalePrice* given *LotArea*:

$$I(SalePrice, LotArea) = H(SalePrice) - H(SalePrice \mid LotArea) \tag{5.6}$$

More generally, the **Mutual Information** *I* between variables *X* and *Y* is defined by:

$$I(X,Y) = H(X) - H(X \mid Y) \tag{5.7}$$

which is equivalent to:

$$I(X,Y) = \sum_{x \in X} \sum_{y \in Y} p(x,y) \log_2 \frac{p(x,y)}{p(x)p(y)} \tag{5.8}$$

and furthermore also equivalent to:

$$I(X,Y) = \sum_{y \in Y} p(y) \sum_{x \in X} p(x \mid y) \log_2 \frac{p(x \mid y)}{p(x)} \tag{5.9}$$

This allows us to compute the **Mutual Information** between a target variable and any possible predictors. As a result, we can find out which predictor provides the maximum information gain and, thus, has the greatest predictive importance.

Chapter 5

Parameter Estimation

Now we see the real benefit of bringing all variables as nodes into BayesiaLab. All the terms of the equation (5.7) can be easily computed with BayesiaLab once we have a fully specified network.

We start with a pair of nodes, namely *Neighborhood* and *SalePrice*. As opposed to *LotArea*, which is a discretized **Continuous** variable, *Neighborhood* is categorical, and, as such, it has been automatically treated as **Discrete** in BayesiaLab. This is the reason the node corresponding to *Neighborhood* has a solid border. We now add an arc between these two nodes, as illustrated in Figure 5.25, so as to explicitly represent the dependency between them.

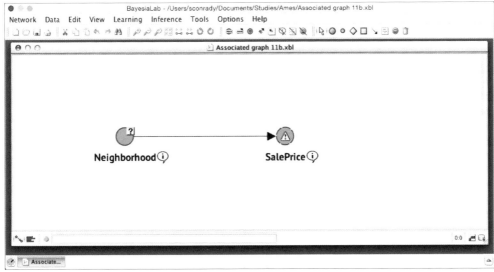

Figure 5.25

The yellow warning triangle (⚠) reminds us that the **Conditional Probability Table (CPT)** of *SalePrice* given *Neighborhood* has not been defined yet. In Chapter 4 we defined the **CPT** from existing knowledge. Here, on the other hand, as we have an associated database, BayesiaLab can use it to estimate the **CPT** by using **Maximum Likelihood**, i.e. BayesiaLab "counts" the (co-)occurrences of the states of the variables in our data. For reference, Figure 5.26 shows the first 10 records of the variables *SalePrice* and *Neighborhood* from the Ames database.

SalePrice	Neighborhood
<=225000	Old Town
<=150000	College Creek
>300000	College Creek
<=150000	Sawyer
<=225000	Gilbert
<=150000	Mitchell
<=150000	Old Town
<=225000	Gilbert
<=150000	Sawyer
<=225000	Gilbert

Figure 5.26

Counting all records, we obtain the marginal count of each state of *Neighborhood* (Figure 5.27).

Neighborhood	Count of Neighborhood
Bloomington Heights	28
Bluestem	10
Briardale	30
Brookside	108
Clear Creek	44
College Creek	267
Crawford	103
Edwards	194
Gilbert	165
Green Hills	2
Greens	8
Iowa DOT and Rail Road	93
Landmark	1
Meadow Village	37
Mitchell	114
Northpark Villa	23
Northridge	71
Northridge Heights	166
Northwest Ames	131
Old Town	239
Sawyer	151
Sawyer West	125
Somerset	182
South & West of Iowa State University	48
Stone Brook	51
Timberland	72
Veenker	24

Figure 5.27

Given that our Bayesian network structure says that *Neighborhood* is the parent node of *SalePrice*, we now count the states of *SalePrice* conditional on *Neighborhood*. This is simply a cross-tabulation (Figure 5.28).

Neighborhood	<=75000	<=150000	<=225000	<=300000	>300000
Bloomington Heights	0	0	23	5	0
Bluestem	0	6	4	0	0
Briardale	0	30	0	0	0
Brookside	8	82	18	0	0
Clear Creek	0	6	20	16	2
College Creek	0	56	139	62	10
Crawford	0	24	45	24	10
Edwards	8	148	28	7	3
Gilbert	0	4	143	15	3
Green Hills	0	0	0	1	1
Greens	0	0	8	0	0
Iowa DOT and Rail Road	21	65	7	0	0
Landmark	0	1	0	0	0
Meadow Village	3	33	1	0	0
Mitchell	0	53	50	11	0
Northpark Villa	0	22	1	0	0
Northridge	0	0	2	32	37
Northridge Heights	0	0	31	44	91
Northwest Ames	0	14	95	21	1
Old Town	16	184	33	4	2
Sawyer	2	112	37	0	0
Sawyer West	1	33	67	21	3
Somerset	0	10	81	72	19
South & West of Iowa State University	1	32	15	0	0
Stone Brook	0	1	12	11	27
Timberland	0	2	31	24	15
Veenker	0	1	9	10	4

Count of Occurrences of SalePrice Given Neighborhood

Figure 5.28

Once we translate these counts into probabilities (by normalizing by the total number of occurrences for each row in the table), this table becomes a **CPT**. Together, the network structure (qualitative) and the **CPTs** (quantitative) make up the Bayesian network, as shown in the conceptual illustration in Figure 5.29.

Neighborhood	Marginal Probabilities of Neighborhood
Bloomington Heights	16.1%
Bluestem	0.3%
Briardale	1.0%
Brookside	3.7%
Clear Creek	1.5%
College Creek	9.1%
Crawford	3.5%
Edwards	6.6%
Gilbert	5.6%
Green Hills	0.1%
Greens	0.3%
Iowa DOT and RR	3.2%
Landmark	0.0%
Meadow Village	1.3%
Mitchell	3.9%
Northpark Villa	0.8%
Northridge	2.4%
Northridge Heights	5.7%
Northwest Ames	4.5%
Old Town	8.2%
Sawyer	5.2%
Sawyer West	4.3%
Somerset	6.2%
South & West of ISU	1.6%
Stone Brook	1.7%
Timberland	2.5%
Veenker	0.8%

Conditional Probabilities of SalePrice Given Neighborhood

Neighborhood	<=75000	<=150000	<=225000	<=300000	>300000
Bloomington Heights	0.0%	0.0%	82.1%	17.9%	0.0%
Bluestem	0.0%	60.0%	40.0%	0.0%	0.0%
Briardale	0.0%	100.0%	0.0%	0.0%	0.0%
Brookside	7.4%	75.9%	16.7%	0.0%	0.0%
Clear Creek	0.0%	13.6%	45.5%	36.4%	4.5%
College Creek	0.0%	21.0%	52.1%	23.2%	3.7%
Crawford	0.0%	23.3%	43.7%	23.3%	9.7%
Edwards	4.1%	76.3%	14.4%	3.6%	1.5%
Gilbert	0.0%	2.4%	86.7%	9.1%	1.8%
Green Hills	0.0%	0.0%	0.0%	50.0%	50.0%
Greens	0.0%	0.0%	100.0%	0.0%	0.0%
Iowa DOT and RR	22.6%	69.9%	7.5%	0.0%	0.0%
Landmark	0.0%	100.0%	0.0%	0.0%	0.0%
Meadow Village	8.1%	89.2%	2.7%	0.0%	0.0%
Mitchell	0.0%	46.5%	43.9%	9.6%	0.0%
Northpark Villa	0.0%	95.7%	4.3%	0.0%	0.0%
Northridge	0.0%	0.0%	2.8%	45.1%	52.1%
Northridge Heights	0.0%	0.0%	18.7%	26.5%	54.8%
Northwest Ames	0.0%	10.7%	72.5%	16.0%	0.8%
Old Town	6.7%	77.0%	13.8%	1.7%	0.8%
Sawyer	1.3%	74.2%	24.5%	0.0%	0.0%
Sawyer West	0.8%	26.4%	53.6%	16.8%	2.4%
Somerset	0.0%	5.5%	44.5%	39.6%	10.4%
South & West of ISU	2.1%	66.7%	31.3%	0.0%	0.0%
Stone Brook	0.0%	2.0%	23.5%	21.6%	52.9%
Timberland	0.0%	2.8%	43.1%	33.3%	20.8%
Veenker	0.0%	4.2%	37.5%	41.7%	16.7%

Figure 5.29

In practice, however, we do not need to bother with these individual steps. Rather, BayesiaLab can automatically learn all marginal and conditional probabilities from the associated database. To perform this task, we select **Learning > Parameter Estimation** (Figure 5.30).

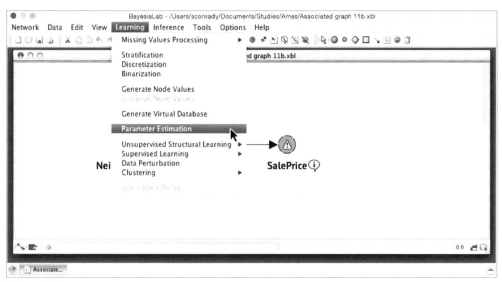

Figure 5.30

Upon completing the **Parameter Estimation**, the warning triangle (⚠) has disappeared, and we can verify the results by double-clicking *SalePrice* to open the **Node Editor**. Under the tab **Probability Distribution** and sub-tab **Probabilistic**, we can see the probabilities of the states of *SalePrice* given *Neighborhood* (Figure 5.31). The **CPT** presented in the **Node Editor** is indeed identical to the table calculated with the spreadsheet (Figure 5.29).

Neighborhood	<=75000	<=150000	<=225000	<=300000	>300000
Bloomington Heights	0.002	0.003	83.700	16.292	0.002
Bluestem	0.006	63.142	36.840	0.006	0.006
Briardale	0.002	99.992	0.002	0.002	0.002
Brookside	5.944	78.898	15.157	0.001	0.001
Clear Creek	0.002	15.971	46.583	33.365	4.080
College Creek	0.000	23.944	52.010	20.770	3.276
Crawford	0.001	26.712	43.830	20.929	8.528
Edwards	3.313	79.371	13.141	2.942	1.233
Gilbert	0.000	2.794	87.393	8.208	1.606
Green Hills	0.037	0.047	0.042	50.495	49.379
Greens	0.008	0.010	99.965	0.008	0.008
Iowa DOT and Rail Road	18.566	74.419	7.014	0.001	0.001
Landmark	0.057	99.764	0.064	0.058	0.057
Meadow Village	6.402	91.176	2.420	0.002	0.002
Mitchell	0.001	50.296	41.524	8.179	0.001
Northpark Villa	0.003	96.165	3.828	0.003	0.002
Northridge	0.001	0.001	3.173	45.443	51.381
Northridge Heights	0.000	0.001	20.658	26.251	53.090
Northwest Ames	0.001	12.243	72.698	14.388	0.670
Old Town	5.374	80.034	12.562	1.363	0.667
Sawyer	1.059	76.751	22.189	0.000	0.000
Sawyer West	0.695	29.688	52.747	14.802	2.068
Somerset	0.000	6.586	46.679	37.148	9.587
South & West of Iowa State University	1.683	69.715	28.599	0.001	0.001
Stone Brook	0.001	2.452	25.734	21.120	50.692
Timberland	0.001	3.367	45.651	31.642	19.340
Veenker	0.003	5.060	39.824	39.616	15.498

Figure 5.31

This model now provides the basis for computing the **Mutual information** between *Neighborhood* and *SalePrice*. BayesiaLab computes **Mutual Information** on demand and can display its value in numerous ways. For instance, from within the **Validation Mode** (≣ or [F5]), we can select **Analysis > Visual > Arcs' Mutual Information**.

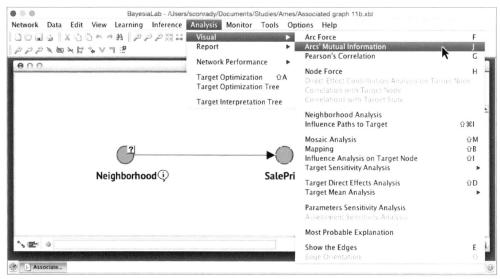

Figure 5.32

The value of **Mutual Information** is now represented graphically in the thickness of the arc. Given that we only have a single arc in this network, this does not give us much insight. So, we click **Display Arc Comments** (▦) to show the numerical values (Figure 5.33).

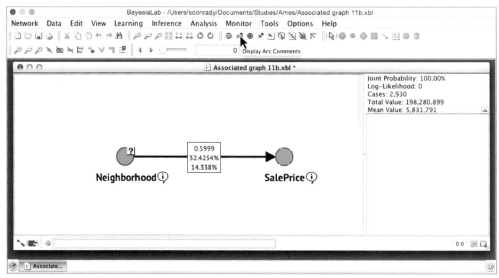

Figure 5.33

Chapter 5

The top number in the yellow box shows the actual **MI** value, i.e. 0.5999 bits (Figure 5.25). We should also point out that **Mutual Information** is a symmetric measure. As such, the amount of **MI** that *Neighborhood* provides on *SalePrice* is the same as the amount of **MI** that *SalePrice* provides with regard to *Neighborhood*. This means that knowing the *SalePrice* reduces the uncertainty with regard to *Neighborhood*, even though that may not be of interest.

```
0.5999
32.4254%
14.338%
```
Figure 5.34

Without context, however, the value of **Mutual Information** number is not meaningful. Hence, BayesiaLab provides two additional measures, shown in red and in blue. The blue number shows the **Relative Mutual Information** with regard to the child node, *SalePrice*, which gives us a sense by how much the entropy of *SalePrice* was reduced. Previously, we computed the marginal entropy of *SalePrice* to be 1.85. Dividing the **Mutual Information** by the marginal entropy of *SalePrice* gives us a sense of how much our uncertainty is reduced:

$$\frac{0.5999}{1.85} = 0.3243 = 32.43\% \tag{5.10}$$

Conversely, the red number shows the **Relative Mutual Information** with regard to the parent node, *Neighborhood*. Here, we divide the **Mutual Information**, which is the same in both directions, by the marginal entropy of *Neighborhood*:

$$\frac{0.5999}{4.1839} = 0.1434 = 14.34\% \tag{5.11}$$

This means that by knowing *Neighborhood*, we reduce our uncertainty regarding *SalePrice* by 32% on average. By knowing *SalePrice*, we reduce our uncertainty regarding *Neighborhood* by 14% on average. These values are readily interpretable. However, we need to know this for all nodes to determine which node is most important.

Naive Bayes Network

Rather than computing the relationships individually for each pair of nodes, we ask BayesiaLab to estimate a **Naive Bayes** network. A **Naive Bayes** structure is a network with only one parent, the **Target Node**, i.e. the only arcs in the graph are those direct-

> The **Naive Bayes** network is perhaps the most commonly used Bayesian network, presumably due to its simplicity. As a result, we find it implemented in many software packages. The so-called Bayesian anti-spam systems are based on this model. It is important to point out, however, that the **Naive Bayes** network is simply the first step towards embracing the Bayesian network paradigm. Therefore, we only show this type of network as an expository aid, rather than proposing to use it as a primary model.

ly connecting the **Target Node** to a set of nodes. By designating *SalePrice* as the **Target Node**, way we can automatically compute its **Mutual Information** with all other available nodes. First, we right-click on the *SalePrice* node to bring up the **Contextual Menu**, from which we then select **Set as Target Node** (Figure 5.35). Alternatively, we can double-click the node while pressing $\boxed{\text{T}}$.

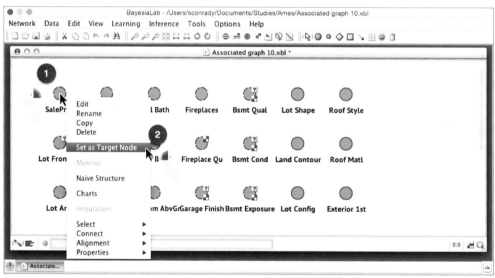

Figure 5.35

A pop-up window requires us to confirm the selection of the **Target Node**. We also have the option of setting a **Target State** of the **Target Node** via the drop-down menu. In our case, the **Target State** is not relevant, and we can leave it at the default value (Figure 5.36). The **Target State** is only useful for certain analysis functions but not for machine learning a structure.

Chapter 5

Figure 5.36

The special status of the **Target Node** is highlighted by the bullseye symbol (◎). We can now proceed to learn the **Naive Bayes** network. From the main menu, we select **Learning > Supervised Learning > Naive Bayes** (Figure 5.37).

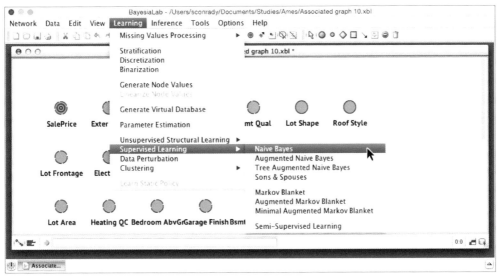

Figure 5.37

Strictly speaking, we are specifying and estimating this network, rather than learning it. Indeed, the "naive" design of the network fully defines its structure (Figure 5.38).

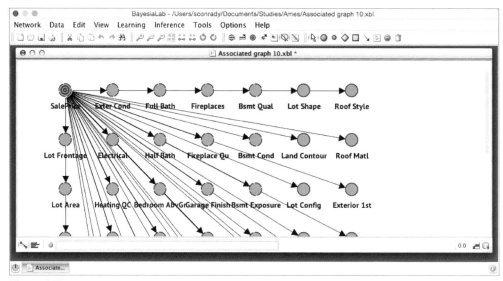

Figure 5.38

Now we have the network in place that allows computing the **Mutual Information** for all nodes. We switch to **Validation Mode** (≝ or F5) and select **Analysis > Visual > Arcs' Mutual Information**.

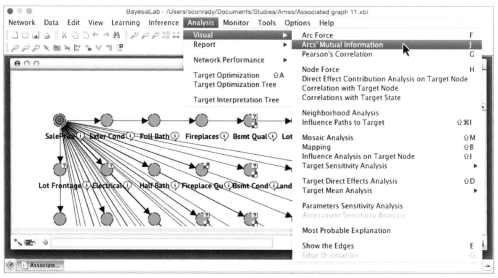

Figure 5.39

The different levels of **Mutual Information** are now reflected in the thickness of the arcs (Figure 5.40).

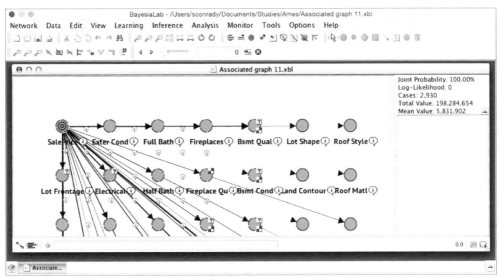

Figure 5.40

However, given the grid layout of the nodes and the overlapping arcs, it is difficult to establish a rank order of the nodes in terms of **Mutual Information**. To address this, we can adjust the layout and select **View > Layout > Radial Layout** (Figure 5.41).

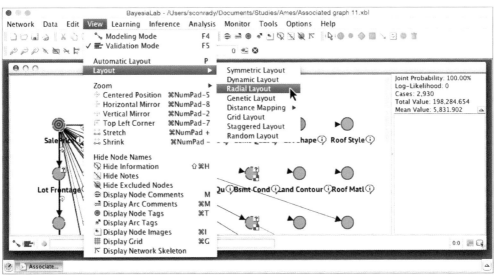

Figure 5.41

This generates a circular arrangement of all nodes with the **Target Node**, *SalePrice*, in the center. By repeatedly clicking the **Stretch** icon ( ), we expand the network to make it fit the available screen space (Figure 5.42). To improve legibility further, we click the **Hide Information** icon in the menu bar ( ).

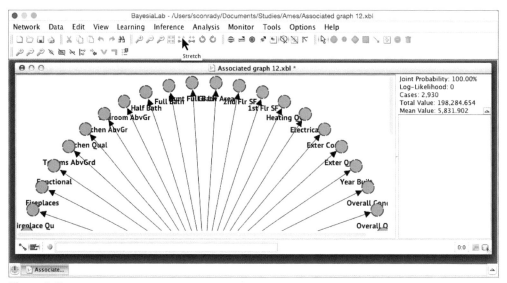

Figure 5.42

By running **Radial Layout** while the visualization of **Arcs' Mutual Information** is still active, the nodes are ordered clockwise from strongest to weakest. To make it easier to see the details of the network, we show it as a standalone graphic in Figure 5.43

Chapter 5

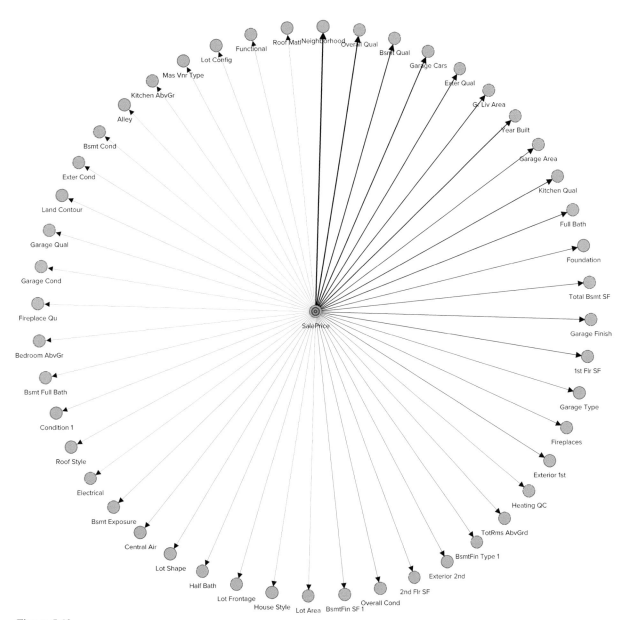

Figure 5.43

The illustration shows that *Neighborhood* provides the highest amount of **Mutual Information**, and, at the opposite end of the range, *RoofMtl (Roof Material)* the least. As an alternative to this visualization, we can run a report: **Analysis > Report > Relationship Analysis**. The resulting **Relationship Analysis** is shown in Figure 5.44.

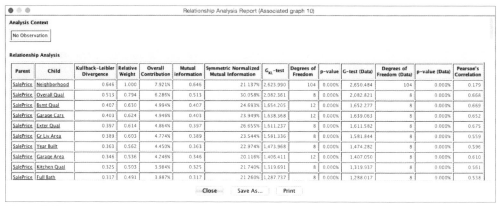

Figure 5.44

Mutual Information vs. Correlation

Wouldn't this report look just the same if it were computed based on correlation? In fact, the rightmost column in the **Relationship Analysis Report** shows **Pearson's Correlation** for reference. As we can see, the order would be different if we were to choose the **Pearson's Correlation** as the main metric. So, have we gained anything over correlation? One of the key advantages of **Mutual Information** is that it can be computed—and interpreted—between numerical and categorical variables, without any variable transformation. For instance, we can easily compute the **Mutual Information**, such as between the *Neighborhood* and *SalePrice*. The question regarding the most important predictive variable can now be answered. It is *Neighborhood*.

Now that we have established the central role of entropy and **Mutual Information**, we can apply these concepts in the next chapters for machine learning and network analysis.

Chapter 6

6. Supervised Learning

In Chapter 4 we defined the qualitative and the quantitative part of a Bayesian networks network from existing (human) knowledge. Chapter 5 described how we can define the qualitative part of a Bayesian network manually and then use data to estimate the quantitative part. In this chapter, we use BayesiaLab for generating both the structure and the parameters of a network automatically from data. This means we introduce machine learning for building Bayesian networks. The only guidance (or constraint) we provide is that we define the variable of interest, i.e. the target of the machine-learning process. Hence, we speak of **Supervised Learning** (in Chapter 7 we will remove that constraint as well and perform **Unsupervised Learning**).

The objective of what we call **Supervised Learning** is no different from that of predictive modeling. We simply wish to find regularities (a model) between the target variable and potential predictors from observations (e.g. historical data). Such a model will subsequently allow us to infer a distribution of the target variable from new observations. If the target variable is **Continuous**, the predicted distribution produces an expected value. For a **Discrete** target variable, we perform classification. The latter will be the objective of the example in this chapter.

Example: Tumor Classification

Given the sheer amount of medical knowledge in existence today, plus advances in artificial intelligence, so-called medical expert systems have emerged, which are meant to support physicians in performing medical diagnoses. In this context, several papers by Wolberg, Street, Heisey, and Managasarian became much-cited examples. For instance, Mangasarian, Street and Wolberg (1995) proposed an automated method for the classification of Fine Needle Aspirates through imaging processing and machine learning with the objective of achieving a greater accuracy in distinguishing between malignant and benign cells for the diagnosis of breast cancer. At the time of their study, the practice of visual inspection of FNA yielded inconsistent diagnostic accuracy. The proposed new approach would increase this accuracy reliably to over 95%.

This research was quickly translated into clinical practice and has since been applied with continued success.

As part of their studies in the late 1980s and 1990s, the research team generated what became known as the Wisconsin Breast Cancer Database, which contains measurements of hundreds of FNA samples and the associated diagnoses. This database has been extensively studied, even outside the medical field. Statisticians and computer scientists have proposed a wide range of techniques for this classification problem and have continuously raised the benchmark for predictive performance.

The objective of this chapter is to show how Bayesian networks, in conjunction with machine learning, can be used for classification. Furthermore, we wish to illustrate how Bayesian networks can help researchers generate a deeper understanding of the underlying problem domain. Beyond merely producing predictions, we can use Bayesian networks to precisely quantify the importance of individual variables and employ BayesiaLab to help identify the most efficient path towards diagnosis.

To provide further background regarding this example, we quote Mangasarian et al. (1994):

> *"Most breast cancers are detected by the patient as a lump in the breast. The majority of breast lumps are benign, so it is the physician's responsibility to diagnose breast cancer, that is, to distinguish benign lumps from malignant ones. There are three available methods for diagnosing breast cancer: mammography, FNA with visual interpretation and surgical biopsy. The reported sensitivity (i.e., ability to correctly diagnose cancer when the disease is present) of mammography varies from 68% to 79%, of FNA with visual interpretation from 65% to 98%, and of surgical biopsy close to 100%.*
>
> *Therefore mammography lacks sensitivity, FNA sensitivity varies widely, and surgical biopsy, although accurate, is invasive, time consuming and costly. The goal of the diagnostic aspect of our research is to develop a relatively objective*

system that diagnoses FNAs with an accuracy that approaches the best achieved visually."

Data: Wisconsin Breast Cancer Database

The Wisconsin Breast Cancer Database was created through the clinical work of Dr. William H. Wolberg at the University of Wisconsin Hospitals in Madison. As of 1992, Dr. Wolberg had collected 699 instances of patient diagnoses in this database, consisting of two classes: 458 benign cases (65.5%) and 241 malignant cases (34.5%). The following eleven attributes are included in the database:

1. Sample code number
2. Clump Thickness (1–10)
3. Uniformity of Cell Size (1–10)
4. Uniformity of Cell Shape (1–10)
5. Marginal Adhesion (1–10)
6. Single Epithelial Cell Size (1–10)
7. Bare Nuclei (1–10)
8. Bland Chromatin (1–10)
9. Normal Nucleoli (1–10)
10. Mitoses (1–10)
11. Class (benign/malignant)

Attributes #2 through #10 were computed from digital images of fine needle aspirates of breast masses. These features describe the characteristics of the cell nuclei in the image. Attribute #11, *Class*, was established via subsequent biopsies or long-term monitoring of the tumor. We will not go into detail here regarding the definition of the attributes and their measurement. Rather, we refer the reader to papers referenced in the bibliography. The Wisconsin Breast Cancer Database is available to any interested researcher from the UC Irvine Machine Learning Repository. We use this database in its original format without any further transformation, so our results can be directly compared to dozens of methods that have been developed since the original study.

Data Import Wizard

Our modeling process begins with importing the database, which is formatted as a text file with comma-separated values. We start the **Data Import Wizard** with **Data > Open Data Source > Text File** (Figure 6.1).

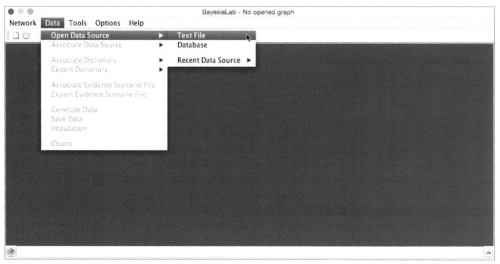

Figure 6.1

Next, we select the file "breast-cancer-wisconsin-data.csv", a comma-delimited, flat text file.[1] The **Data Import Wizard** then guides us through the required steps. In the first dialogue box of the **Data Import Wizard**, we click on **Define Typing** and specify that we wish to set aside a **Test Set** from the database (Figure 6.2).

Figure 6.2

We arbitrarily select the first 139 observations as a custom **Test Set**, and, consequently, the remaining cases will serve as our **Learning Set** (Figure 6.3).

---

1 The Wisconsin Breast Cancer Database is available for download from the Bayesia website: www.bayesia.us/wbcd

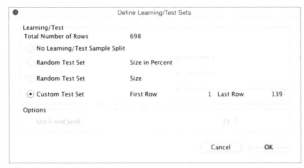

Figure 6.3

In the next step, the **Data Import Wizard** suggests the data type for each variable. Attributes #2 through #10 are identified as continuous variables, and *Class* is read as a **Discrete** variable (Figure 6.4). Only for the first variable, *Sample code number*, we have to specify **Row Identifier**, so it is not mistaken for a continuous predictor variable.

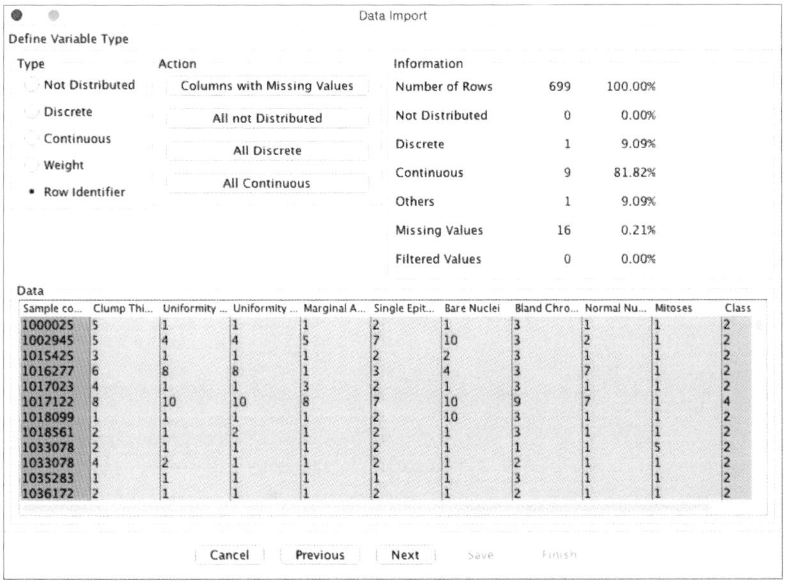

Figure 6.4

In the next step, the **Information Panel** reports that we have a total of 16 missing values in the entire database. We can also see that the column *Bare Nuclei* is labeled with a small question mark (?), which indicates the presence of missing values in this particular column. Therefore, we must specify the type of **Missing Values Imputation**. Given the small size of the dataset, and the small number of missing values, we will choose the **Structural EM** method (Figure 6.5).

▶ Chapter 9. Missing Values Processing, p. 289.

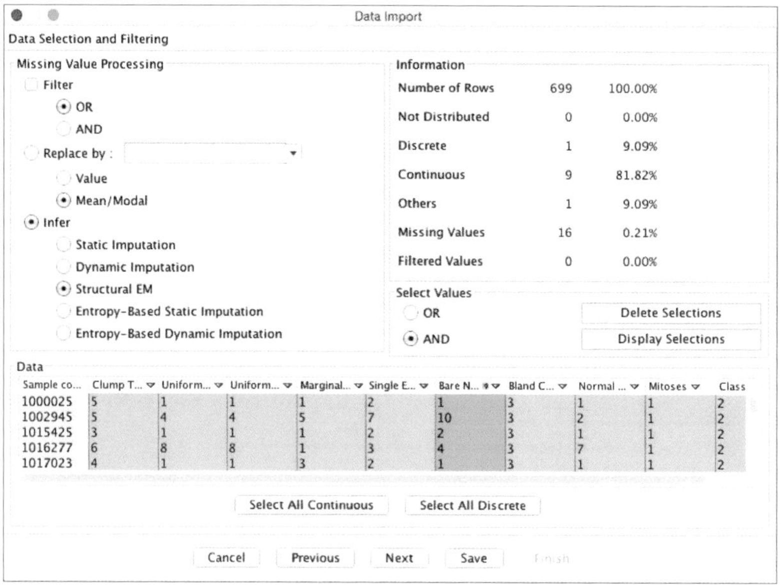

Figure 6.5

A central element of the data import process is the discretization of **Continuous** variables. Even though we could select a specific discretization method for each **Continuous** variables, we choose to apply the same algorithm to all. So, on the next screen we click **Select All Continuous** to apply the same discretization algorithm across all **Continuous** variables.

▶ Tree in Chapter 5, p. 88.

As the objective of this exercise is classification, we are then in the supervised learning framework where we have a Target node. Thus, we choose the **Tree** algorithm—BayesiaLab's only bivariate discretization algorithm—from the drop-down menu in the **Multiple Discretization** panel (Figure 6.6).

Chapter 6

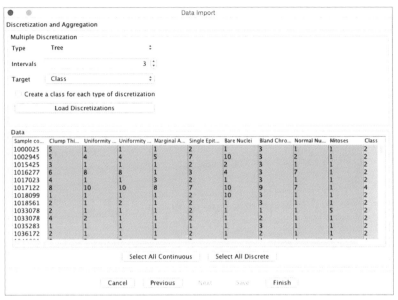

Figure 6.6

Discretization Intervals

▸ Data Discretization in Chapter 7, p. 170.

Bayesian networks are non-parametric probabilistic models. Therefore, there is no hypothesis with regard to the form of the relationships between variables (e.g. linear, quadratic, exponential, etc.). However, this flexibility has a cost, the number of observations necessary to quantify probabilistic relationships is higher than those required in parametric models. We use the heuristic of five observations per probability cell, which implies that the bigger the size of the probability tables, the larger must be the number of observations.

Two parameters affect the size of a probability table: the number of parents and the number of states of the parent and child nodes. A machine-learning algorithm usually determines the number of parents based on the strength of the relationships and the number of available observations. The number of states, however, is our choice, which we can set by means of **Discretization** (for **Continuous** variables) and **Aggregation** (for **Discrete** variables).

We can use our heuristic of five observations per probability cell to help us with the selection of the number of **Discretization Intervals**:

$$StateCount^{ParentCount+1} \times 5 \leq Observations \qquad (6.1)$$

We usually look for an odd number of states to be able to capture non-linear relationships. Given that we have a relatively small learning set of only 560 observations, we should estimate how many parents would be allowed based on this heuristic and a discretization with 3 states:

- No parent: 3×5=15
- One parent: 3×3×5=45
- Two parents: 3×3×3×5=135
- Three parents: 3×3×3×3×5=405
- Four parents: 3×3×3×3×3×5=1,215

Considering a discretization with 5 states, we would obtain:

- No parent: 5×5=25
- One parent: 5×5×5=125
- Two parents: 5×5×5×5=625

By using this heuristic, we hypothesize about the size of the biggest CPT of the to-be-learned Bayesian network and multiply this value by 5. Experience tells us that this is a rather practical heuristic, which typically helps us finding a structure. However, this is by no means a guarantee that we will find a precise quantification of the probabilistic relationships.

Indeed, our heuristic is based on the hypothesis that all the cells of the CPT are equally likely to be sampled. Of course, such an assumption cannot hold as the sampling probability of a cell depends on its probability, i.e. either a marginal probability if the node does not have parents, or, if it does have parents, a joint probability defined by the parent states and the child state.

Given our 560 observations and the scenarios listed above, we select a discretization scheme with a maximum of 3 states. This is a maximum in the sense that the **Tree** discretization algorithm could return 2 states if 3 were not needed, e.g. *Mitoses* in Figure 6.8.

Upon clicking **Finish**, BayesiaLab imports and discretizes the entire database and concludes this process by offering an **Import Report**. To see the obtained **Discretization Intervals**, we click **Yes** to bring up the report.

Figure 6.7

It is interesting to see that all the variables have indeed been discretized with the **Tree** algorithm and that all **Discretization Intervals** are variable-specific. This means that all the variables are marginally dependent on the **Target** (and vice versa). This is promising: the more dependent variables we have, the easier it should be to learn a good model for predicting the **Target**.

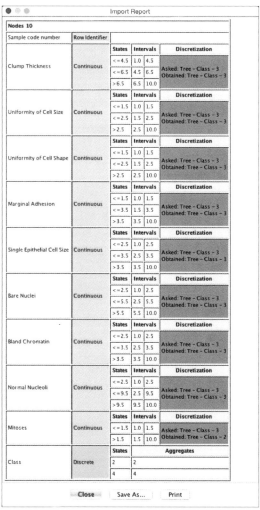

Figure 6.8

Upon closing the **Import Report**, we see a representation of the newly imported database in the form of a fully unconnected Bayesian network in the **Graph Panel** (Figure 6.9).

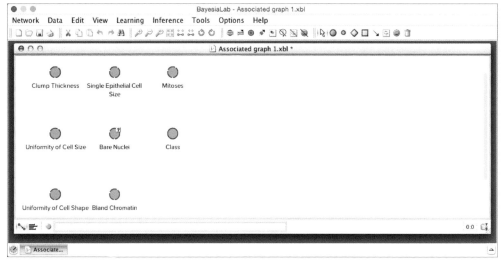

Figure 6.9

The question mark symbol (**?**), which is associated with the *Bare Nuclei* node, indicates that *Bare Nuclei* is the only node with missing values. Hovering over the question mark (**?**) with the cursor, while pressing **I**, shows the number of missing values.

State Names

In the original database for the variable *Class*, codes *2* and *4* represented *benign* and *malignant* respectively. For reading the analysis reports, however, it will be easier to work with a proper **State Name** as opposed to a numeric code. By double-clicking the node *Class*, we open the **Node Editor** and then go to the **State Names** tab. There, we associate the **States** *2* and *4* with new **State Names** (Figure 6.10).

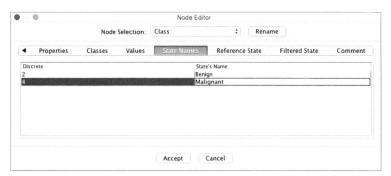

Figure 6.10

Chapter 6

Supervised Learning

Model 1: Markov Blanket

Given our objective of predicting the state of the variable *Class*, i.e. *benign* versus *malignant*, we will define *Class* as the **Target Node**. We need to specify this explicitly, so the subsequent **Supervised Learning** algorithm can focus on the characterization of the **Target Node**, rather than on a representation of the entire joint probability distribution of the learning set. Upon this selection, all **Supervised Learning** algorithms become available under **Learning > Supervised Learning** (Figure 6.11).

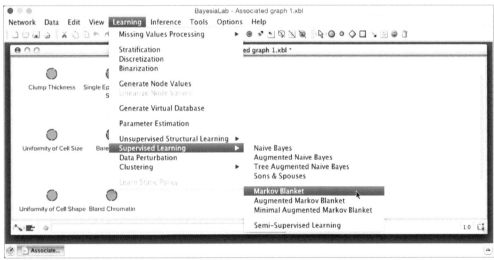

Figure 6.11

Markov Blanket Definition

The Markov Blanket of a node A is the set of nodes composed of A's parents, its children, and its children's other parents (=spouses). The **Markov Blanket** of the node A contains all the nodes that, if we know their states, i.e. we have hard evidence for these nodes, will shield the node A from the rest of the network, i.e. make A independent of all the other nodes given its Markov Blanket (Figure 6.12).

This means that the **Markov Blanket** of a node A is the only knowledge needed to predict the behavior of that node. Learning a **Markov Blanket** selects the most relevant predictor nodes, which is particularly helpful when there is a large number of variables in a dataset. As a result, this can serve as a highly-efficient variable selection method in preparation for other types of modeling, e.g. neural networks.

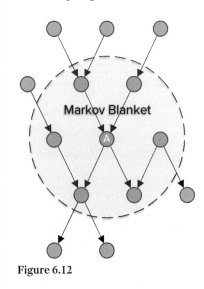

Figure 6.12

Upon learning the **Markov Blanket** for *Class*, and after having applied the **Automatic Layout** (P), the resulting Bayesian network looks as shown in Figure 6.13.

Chapter 6

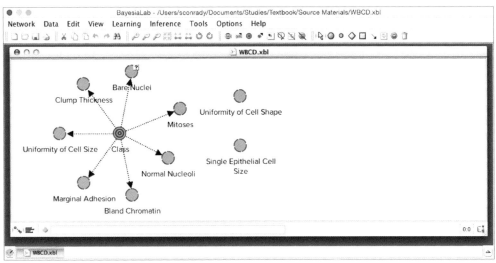

Figure 6.13

We can see that the obtained network is a **Naive** structure on a subset of nodes. This means that *Class* has a direct probabilistic relationship with all nodes except *Uniformity of Cell Shape* and *Single Epithelial Cell Size*, which are both disconnected. The lack of their connection with the **Target Node** implies that these nodes are independent of the **Target Node** given the nodes in the **Markov Blanket**.

Beyond distinguishing between predictors (connected nodes) and non-predictors (disconnected nodes), we can further examine the relationship versus the **Target Node** *Class* by highlighting the **Mutual Information** of the arcs connecting the nodes. This function is accessible within the **Validation Mode** (≡ or F5) via **Analysis > Visual > Arcs' Mutual Information** (Figure 6.14).

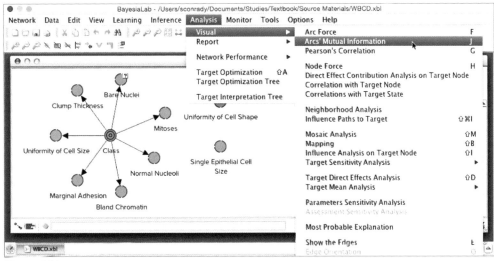

Figure 6.14

Applying a **Radial Layout** (**View > Layout > Radial Layout**) furthermore orders the nodes and arcs clockwise according to their **Mutual Information** with the **Target Node** (Figure 6.15).

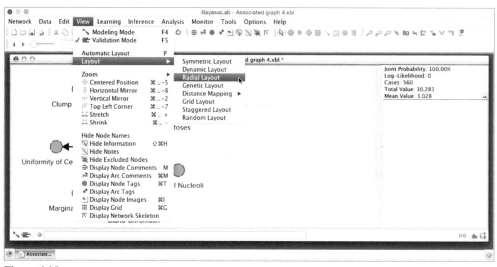

Figure 6.15

▶ Mutual Information in Chapter 5, p. 98.

Finally, we select **View > Display Arc Comments** (M). This allows us to examine the **Mutual Information** between all nodes and the **Target Node** *Class*, which enables us to gauge the relative importance of the nodes.

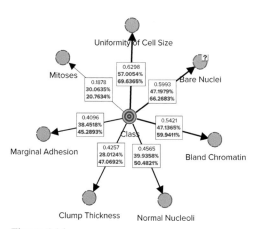

Figure 6.16

Model 1: Performance Analysis

As we are not equipped with specific domain knowledge about the nodes, we will not further interpret these relationships but rather run an initial test regarding the

Network Performance. We want to know how well this **Markov Blanket** model can predict the states of the *Class* variable, i.e. *Benign* versus *Malignant*. This test is available via **Analysis > Network Performance > Target** (Figure 6.17).

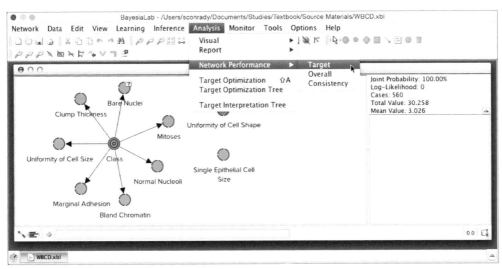

Figure 6.17

As the analysis starts, BayesiaLab prompts us to **Define Acceptance Thresholds**. With the given binary **Target Node**, we select **Evaluate All States** and proceed.

Using our previously defined **Test Set** for evaluating our model, we obtain the initial performance results (Figure 6.18), including metrics, such as **Total Precision**, R, R^2, etc.

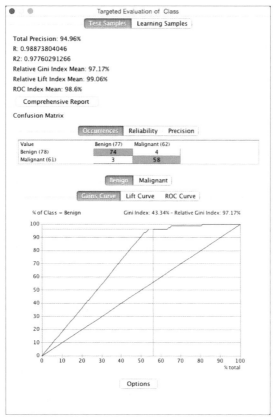

Figure 6.18

In the context of this example, however, the table in the center of the report warrants special attention. For closer examination, we have copied the tables into an annotated spreadsheet (Figure 6.19). Of the 77 *Benign* cases of the test set, 3 were incorrectly identified, which corresponds to a false positive rate of 3.9%. More importantly though, of the 62 *Malignant* cases, 93.55% were identified correctly (true positives) with 4 false negatives. The overall performance can be expressed as the **Total Precision**, which is computed as total number of correct predictions (true positives + true negatives) divided by the total number of cases in the Test Set, i.e. (58+74)÷139=94.96%.

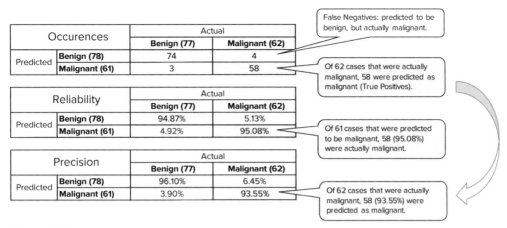

Figure 6.19

K-Folds Cross-Validation

While those results may appear encouraging with regard to the model performance, we need to bear in mind that we arbitrarily selected a **Test Set**, i.e. the first 139 cases in the database.

To mitigate any sampling artifacts that may occur in such a one-off **Test Set**, we can systematically learn networks on a sequence of different subsets and then aggregate the test results. Similarly to the original studies of this topic, we will perform **K-Folds Cross Validation**, which will iteratively select K different **Learning Sets** and **Test Sets** and then, based on those, learn the networks and test their performance. **K-Folds Cross Validation** can then be started via **Tools > Cross Validation > Targeted Evaluation > K-Folds** (Figure 6.20).

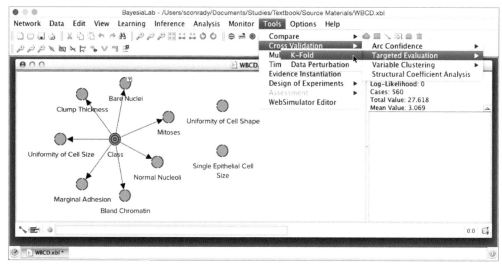

Figure 6.20

We use the same learning algorithm as before, i.e. the **Markov Blanket**, and we choose K=10 as the number of sub-samples to be analyzed. Of the total dataset of 699 cases, each of the ten iterations (folds) will create a **Test Set** of 69 randomly drawn samples, and use the remaining 630 as the **Learning Set** (Figure 6.21). This means that BayesiaLab learns one network per **Learning Set** and then tests the performance on the respective **Test Set**. It is important to ensure that **Shuffle Samples** is checked.[2]

Figure 6.21

---

2   Unchecking **Shuffle Samples** will maintain the original order of the records in the generated folds. This is necessary for datasets with a temporal order, i.e. time series data.

The summary, including the synthesized results, is shown in Figure 6.22. These results confirm a good performance of this model. The **Total Precision** is 96.7%, with a false negative rate of 3.7%. This means 9 of the cases were predicted as *Benign*, while they were actually *Malignant*.

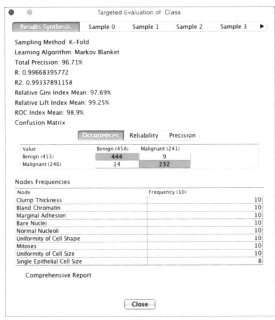

Figure 6.22

Clicking **Comprehensive Report** produces a summary with additional analysis options (Figure 6.23).

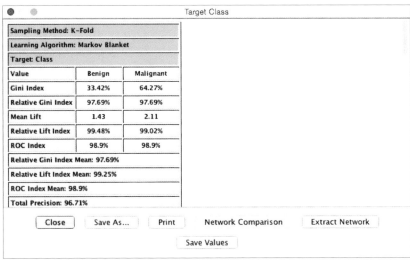

Figure 6.23

For instance, the report can be saved in HTML format, which is convenient for subsequently editing the report as a spreadsheet (Figure 6.24).

Sampling Method: K-Fold		
Learning Algorithm: Markov Blanket		
Target: Class		
Value	Benign	Malignant
Gini Index	33.42%	64.27%
Relative Gini Index	97.69%	97.69%
Mean Lift	1.43	2.11
Relative Lift Index	99.48%	99.02%
ROC Index	98.90%	98.90%
Relative Gini Index Mean: 97.69%		
Relative Lift Index Mean: 99.25%		
ROC Index Mean: 98.9%		
Total Precision: 96.71%		
R: 0.99668395772		
R2: 0.99337891158		
Occurrences		
Value	Benign (458)	Malignant (241)
Benign (453)	444	9
Malignant (246)	14	232
Reliability		
Value	Benign (458)	Malignant (241)
Benign (453)	98.01%	1.99%
Malignant (246)	5.69%	94.31%
Precision		
Value	Benign (458)	Malignant (241)
Benign (453)	96.94%	3.73%
Malignant (246)	3.06%	96.27%

Figure 6.24

To understand what exactly is happening during **K-Folds Cross-Validation**, it is helpful to click **Network Comparison** (Figure 6.23), which initially brings up a view of the **Synthesis Structure** of all the networks learned during the **Cross-Validation**, shown in Figure 6.25 (a). Black arcs in (a) indicate the arcs that were present in the **Reference Structure,** which is displayed in Figure 6.25 (b). This is the network that was learned on the basis on the originally defined **Learning/Test Set.** The thickness of the arcs in (a) reflects how often these links were found in the course of the **Cross-Validation.** The blue-colored arcs indicate that those links were found in some folds, but that they are not part of the **Reference Network.** Their thickness also depends on the number of folds in which these arcs were added.

More specifically, we can scroll through all the networks discovered during the **Cross-Validation** using the record selector icons (◁ ▷). After the **Reference Network** (b), we arrive at **Comparison Network 0** (c). This network structure was learned from 2 out of 10 folds. The last panel (d) shows that **Comparison Network 1** was found 8 out of 10 times.

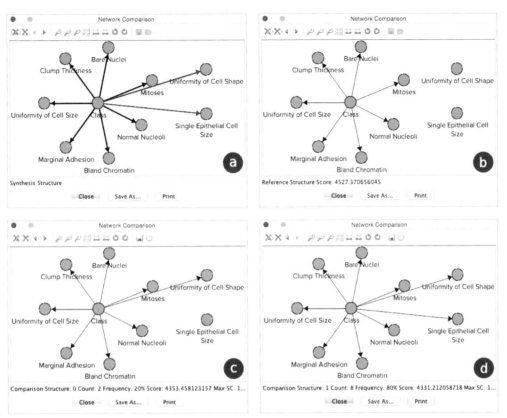

Figure 6.25

This provides numerous important insights. It appears that the first network we learned with the original **Learning Set,** i.e. the **Reference Network**, was not found in any of the 10 learned networks. This is because the learning set we had used in the initial split is smaller than the ones that we used with the 10-fold cross-validation (560 versus 630). The relationships with the two nodes that had been "excluded" are probably the weakest ones.

Model 2: Augmented Markov Blanket

In addition to trying to identify the best network for the **Markov Blanket** algorithm, we also need to consider alternatives within the group of **Supervised Learning** algorithms. BayesiaLab offers an extension to the **Markov Blanket** algorithm, namely the **Augmented Markov Blanket**, which performs an **Unsupervised Learning** algorithm on the nodes in the **Markov Blanket**. This relaxes the constraint of requiring orthogonal child nodes. Thus, it helps identify any influence paths between the predictor variables and potentially improve the predictive performance. Adding such arcs would be similar to automatically creating interaction terms in a regression analysis.

After returning to the **Modeling Mode** (⁕ or [F4]), we start this learning algorithm via **Learning > Supervised Learning > Augmented Markov Blanket** (Figure 6.26). Note, however, that we are still using the original **Learning/Test Sets**. The T symbol (T) on top of the database icon (🗐) reminds us that this database split remains in place.

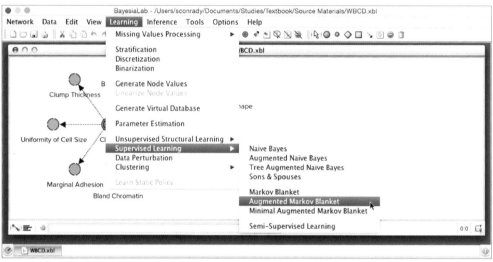

Figure 6.26

As expected, the resulting network is somewhat more complex than the standard **Markov Blanket** (Figure 6.27).

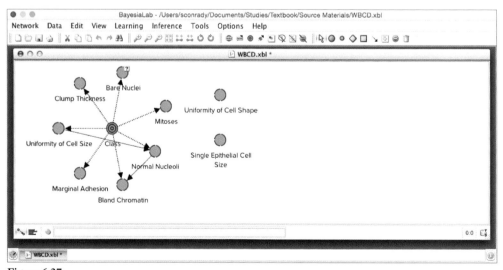

Figure 6.27

Chapter 6

If we save the original **Markov Blanket** and the new **Augmented Markov Blanket** under different file names, we can use **Tools > Compare > Structure** to highlight the differences between both (Figure 6.28). Given that the addition of two arcs is immediately visible, this function may appear as overkill for our example. However, in more complex situations, **Structure Comparison** can be rather helpful, and so we will spell out the details.

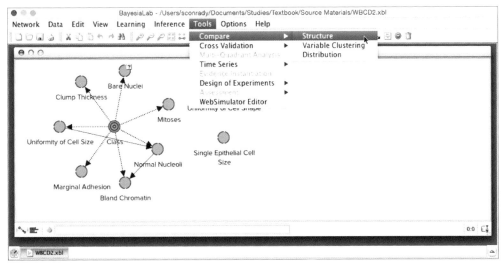

Figure 6.28

We choose the original network and the newly learned network as the **Reference Network** and the **Comparison Network** respectively (Figure 6.29).

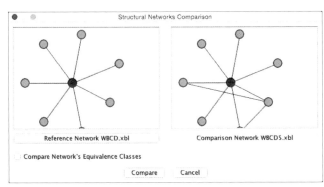

Figure 6.29

Upon selection, a table provides a list of common arcs and of those arcs that have been added to the **Comparison Network**, which was learned with the **Augmented Markov Blanket** algorithm (Figure 6.30):

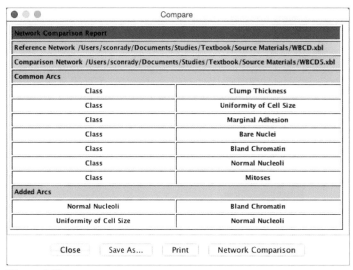

Figure 6.30

Clicking **Charts** provides a visualization of these differences. The arcs that were added by the **Augmented Markov Blanket** are now highlighted in blue. Conversely, had any arcs been deleted, those would be shown in red (Figure 6.31).

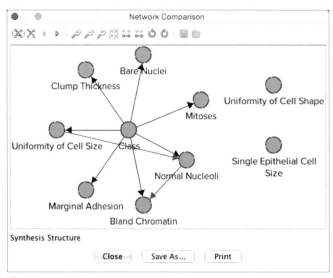

Figure 6.31

Cross-Validation

Given the experience with the **Markov Blanket** model, we perform the **K-Folds Cross-Validation** again: **Tools > Cross-Validation > Targeted Evaluation > K-Fold**.

The steps are identical to what we did earlier, so we move straight to the report (Figure 6.32).

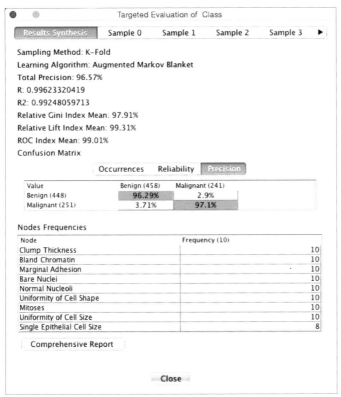

Figure 6.32

At first glance, the predictive performance appears comparable with the **Markov Blanket** model. However, the **Augmented Markov Blanket** model performs slightly better with regard to false negatives, which might be particularly important in the context of cancer diagnostics.

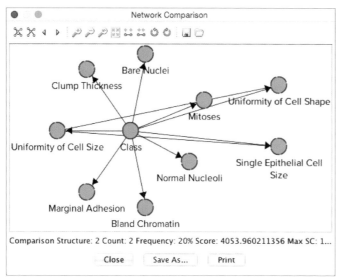

Figure 6.33

For the remainder of this example, we simply remove the **Learning/Test Set** split from the database by right-clicking the database icon (🗄) and selecting **Remove Learning/Test** (Figure 6.34).

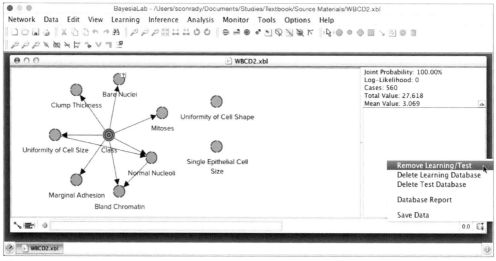

Figure 6.34

Structural Coefficient

Up to this point, the difference in network complexity was only a function of the choice of the learning algorithm and the learning set. We now introduce the **Structural Coefficient (SC)**. This parameter allows changing the internal number of observations and, thus, determines a kind of "significance threshold" for network learning. Consequently, it influences the degree of complexity of the induced networks. The internal number of observations is defined as:

$$N' = \frac{N}{SC} \tag{6.2}$$

where N is the number of samples in the dataset.

By default, **SC** is set to 1, which reliably prevents the learning algorithms from overfitting the model to the data. However, in studies with relatively few observations, the analyst's judgment is needed as to whether a downward adjustment of this parameter can be justified. Reducing **SC** means increasing N', which is like increasing the number of observations in the dataset via resampling.

On the other hand, increasing **SC** beyond 1 means reducing N', which can help manage the complexity of networks learned from large datasets. Conceptually, reducing N' is equivalent to sampling the dataset.

Given the fairly simple network structure of the **Markov Blanket** model, complexity was of no concern. The **Augmented Markov Blanket** is more complex but still very manageable. The question is, could a more complex network provide greater precision without over-fitting? To answer this question, we perform a **Structural Coefficient Analysis**, which generates several metrics that help in making the trade-off between complexity and precision: **Tools > Cross Validation > Structural Coefficient Analysis** (Figure 6.35).

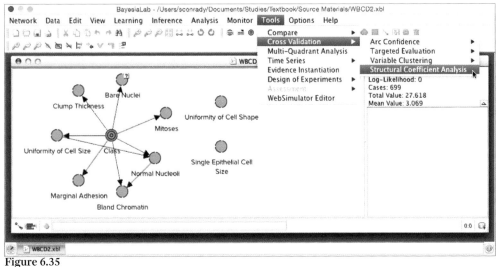

Figure 6.35

BayesiaLab prompts us to specify the range of **SC** value to be examined and the number of iterations to be performed (Figure 6.36). It is worth noting that the minimum **SC** value should not be set to 0, or even close to 0, without careful consideration. An **SC** value of 0 would create a fully connected network, which can take a very long time to learn, depending on the number of variables, or even exceed the memory capacity of the computer running BayesiaLab. Technically, **SC**=0 implies an infinite dataset, which results in all relationships between nodes becoming significant.

Number of Iterations determines the interval steps to be taken within the specified range of the **Structural Coefficient**. Given the relatively light computational load, we choose 25 iterations. With more complex models, we might be more conservative, as each iteration re-learns and re-evaluates the network (Figure 6.36).

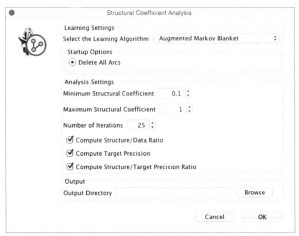

Figure 6.36

The resulting report shows how the network changes as a function of the **Structural Coefficient** (Figure 6.37).

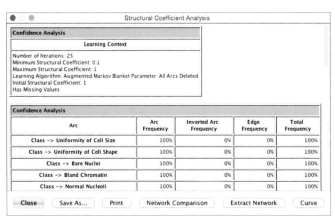

Figure 6.37

140

Our objective is to determine the correct level of network complexity for a reliably high predictive performance while avoiding the over-fitting the data. By clicking **Curve**, we can plot several different metrics for this purpose.

Figure 6.38

Structure/Target Ratio

The **Structure/Target Precision Ratio** (Figure 6.39) is a very helpful measure for making trade-offs between predictive performance versus network complexity. This plot can be best interpreted when following the curve from right to left. Moving to the left along the x-axis lowers the **Structural Coefficient**, which, in turn, results in a more complex **Structure**.

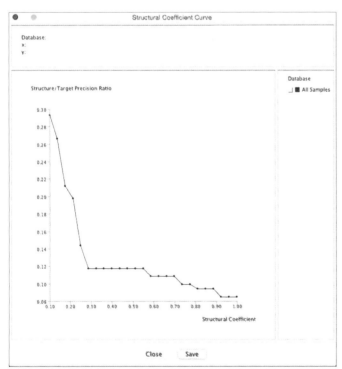

Figure 6.39

It becomes problematic when the **Structure** increases faster than the **Precision**, i.e. we increase complexity without improving **Precision**. Typically, the "elbow" of the

L-shaped curve identifies this critical point. Here, a visual inspection suggests that the "elbow" is just below SC=0.3 (Figure 6.39). The portion of the curve further to the left on the x-axis, i.e. **SC**<0.3, shows that **Structure** is increasing without improving **Precision**, which can be a potential cause of over-fitting. Hence, we conclude that SC=0.3 is a reasonable choice for proceeding further.

Model 2b: Augmented Markov Blanket (SC=0.3)

Given the results from the **Structural Coefficient Analysis**, we now wish to relearn the network with SC=0.3. The **SC** value can be set by right-clicking on the background of the **Graph Panel** and then selecting **Edit Structural Coefficient** from the **Contextual Menu** (Figure 6.40), or via the menu: **Edit > Edit Structural Coefficient**. The **SC** value can then be set with a slider or by typing in a numerical value (Figure 6.41).

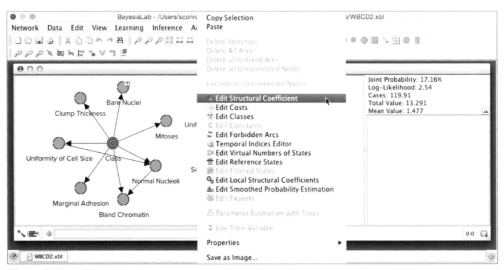

Figure 6.40

Figure 6.41

The **Structural Coefficient** icon (σ) now indicates that we are employing an SC value other than the default of 1. After returning to the **Modeling Mode** (or F4), we relearn the network, using the same **Augmented Markov Blanket** algorithm as before. As expected, this produces a more complex network (Figure

6.42). The key question is, will this increase in complexity improve the performance or perhaps be counterproductive?

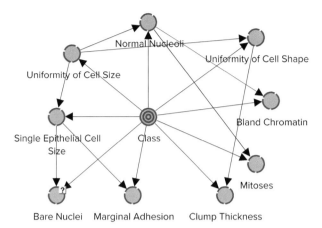

Figure 6.42

Model 2b Performance:

Given that we have removed the **Learning/Test Set** split, we perform **K-Folds Cross-Validation**. We present the results side-by-side in Figure 6.43:

- Model 1 (**Markov Blanket**)
- Model 2a (**Augmented Markov Blanket, SC=1**)
- Model 2b (**Augmented Markov Blanket, SC=0.3**)

Model 1: Markov Blanket (SC=1)			Model 2a: Augmented Markov Blanket (SC=1)			Model 2b: Augmented Markov Blanked (SC=0.3)		
Target: Class			Target: Class			Target: Class		
Value	Benign	Malignant	Value	Benign	Malignant	Value	Benign	Malignant
Gini Index	33.97%	64.55%	Gini Index	34.15%	64.90%	Gini Index	34.25%	65.08%
Relative Gini Index	98.52%	98.52%	Relative Gini Index	99.06%	99.06%	Relative Gini Index	99.33%	99.33%
Mean Lift	1.42	2.05	Mean Lift	1.42	2.05	Mean Lift	1.42	2.06
Relative Lift Index	99.74%	99.19%	Relative Lift Index	99.83%	99.52%	Relative Lift Index	99.88%	99.66%
ROC Index	99.26%	99.26%	ROC Index	99.53%	99.53%	ROC Index	99.66%	99.66%
Relative Gini Index Mean: 98.52%			Relative Gini Index Mean: 99.06%			Relative Gini Index Mean: 99.33%		
Relative Lift Index Mean: 99.47%			Relative Lift Index Mean: 99.68%			Relative Lift Index Mean: 99.77%		
ROC Index Mean: 99.26%			ROC Index Mean: 99.53%			ROC Index Mean: 99.66%		
Total Precision: 96.85%			Total Precision: 97.28%			Total Precision: 97.85%		
R: 0.99757966325			R: 0.99754843972			R: 0.99435420252		
R2: 0.99516518453			R2: 0.99510288959			R2: 0.98874028007		
Occurrences			Occurrences			Occurrences		
Value	Benign (458)	Malignant (241)	Value	Benign (458)	Malignant (241)	Value	Benign (458)	Malignant (241)
Benign (452)	444	8	Benign (453)	446	7	Benign (455)	449	6
Malignant (247)	14	233	Malignant (246)	12	234	Malignant (244)	9	235
Reliability			Reliability			Reliability		
Value	Benign (458)	Malignant (241)	Value	Benign (458)	Malignant (241)	Value	Benign (458)	Malignant (241)
Benign (452)	98.23%	1.77%	Benign (453)	98.45%	1.55%	Benign (455)	98.68%	1.32%
Malignant (247)	5.67%	94.33%	Malignant (246)	4.88%	95.12%	Malignant (244)	3.69%	96.31%
Precision			Precision			Precision		
Value	Benign (458)	Malignant (241)	Value	Benign (458)	Malignant (241)	Value	Benign (458)	Malignant (241)
Benign (452)	96.94%	3.32%	Benign (453)	97.38%	2.90%	Benign (455)	98.03%	2.49%
Malignant (247)	3.06%	96.68%	Malignant (246)	2.62%	97.10%	Malignant (244)	1.97%	97.51%

Figure 6.43

All models reviewed, Model 1 (**Markov Blanket**), Model 2a (**Augmented Markov Blanket, SC=1**), and Model 2b (**Augmented Markov Blanket, SC=0.3**) have performed at fairly similar levels of classification performance. Reestimating these models with more observations could potentially change the results and might more clearly differentiate the classification performance. Given its slight edge in performance, however, we select Model 2b for now to serve as the basis for the next section of the next section, Model Inference.

Model Inference

We now present how the learned **Augment Markov Blanket** model can be applied in practice and used for inference. First, we need to go into **Validation Mode** (≣ or F5) and bring up all the **Monitors** in the **Monitor Panel**. As we have a **Target Node**, we can right-click inside the **Monitor Panel** to activate the corresponding **Contextual Menu** and select Sort > Target Correlation from the **Contextual Menu** (Figure 6.44).

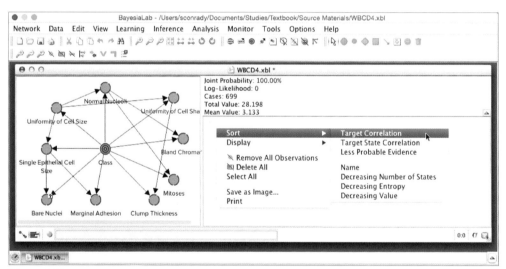

Figure 6.44

Alternatively, we can do the same via **Monitor > Sort > Target Correlation** (Figure 6.45).

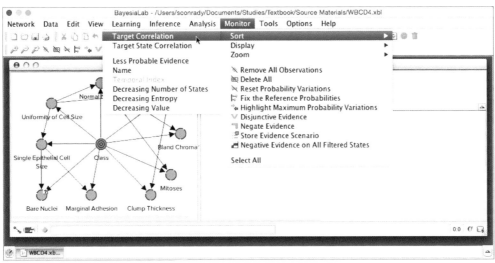

Figure 6.45

Monitors are then automatically created for all the nodes "correlated" with the **Target Node**. The **Monitor** of the **Target Node** is placed first in the **Monitor Panel**, followed by the other **Monitors** in order of their "correlation" with the **Target Node**, from highest to lowest (Figure 6.46).

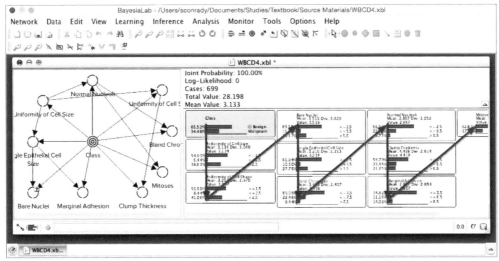

Figure 6.46

As the quotation marks indicate, we use "correlation" not literally in this context. Rather, the sort order is determined by **Mutual Information**.

Interactive Inference

We can use now BayesiaLab to review the individual predictions made based on the model. This feature is called **Interactive Inference**, which can be accessed from the menu via **Inference > Interactive Inference** (Figure 6.47).

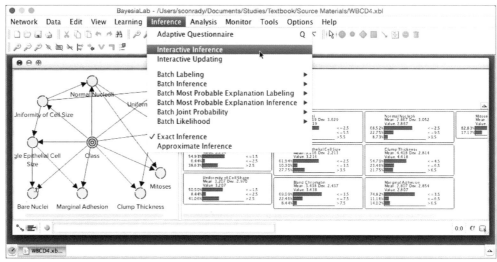

Figure 6.47

The **Navigation Bar** allows scrolling through each record of the database. Record #0 can be seen in Figure 6.48 with all the associated observations highlighted in green. Given these observations, the model predicts a 99.99% probability that *Class* is *Benign* (the **Monitor** of the **Target Node** is highlighted in red). Given this very high probability, calling *Class=Benign* is the rational prediction. For Record #0, it is indeed the correct prediction: The actual value recorded in the dataset is represented by a light blue bar, which signals *Class=Benign* for this case.

Chapter 6

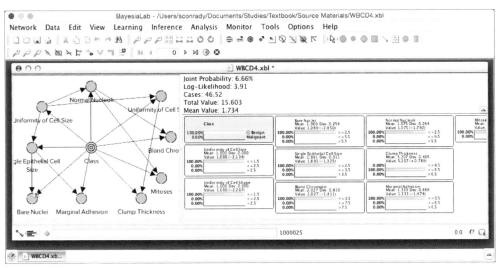

Figure 6.48

Most cases are rather clear-cut, like Record #0, with probabilities for either class around 99% or higher. However, there are a number of exceptions, such as Record #3. Here, the probability of malignancy is approximately 55% (Figure 6.49). Given this probability, the rational prediction is *Class=Malignant*, which, however, turns out to be incorrect.

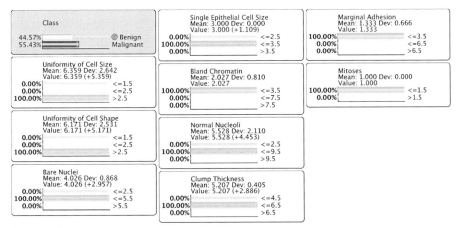

Figure 6.49

Adaptive Questionnaire

In situations in which only individual cases are under review, e.g. when diagnosing a patient, BayesiaLab can provide diagnostic support by means of the **Adaptive**

Questionnaire. The **Adaptive Questionnaire** can be started from the menu via **Inference > Adaptive Questionnaire** (Figure 6.50).

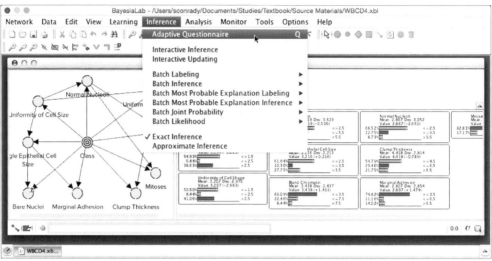

Figure 6.50

For a **Target Node** with more than two states, it can be helpful to specify a **Target State** for the **Adaptive Questionnaire**. Setting the **Target State** allows BayesiaLab to compute **Binary Mutual Information** and then focus on the designated **Target State**. As the **Target Node** *Class* is binary, setting a **Target State** is not useful (Figure 6.51).

Figure 6.51

Costs

Furthermore, we can set the cost of collecting observations via the **Cost Editor**, which can be started via the **Edit Costs** button (Figure 6.52). This is helpful when certain observations are more costly to obtain than others.

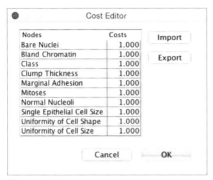

Figure 6.52

Unfortunately, our example is not ideally suited to illustrate this feature, as the FNA attributes are all collected at the same time, rather than consecutively. However, we can imagine that, in other contexts, a physician starts the diagnosis process by collecting easy-to-obtain data, such as blood pressure, before proceeding to more elaborate (and more expensive) diagnostic techniques, such as performing an MRI.

Once the **Adaptive Questionnaire** is started, BayesiaLab presents the **Monitor** of the **Target Node** (red) and its marginal probability. Again, as shown in Figure 6.53, the **Monitors** are automatically sorted in descending order with regard to the **Target Node** by taking into account the **Mutual Information** (or binary mutual information) and the **Cost** of obtaining the information.

Figure 6.53

This means that the ideal first piece of evidence is *Uniformity of Cell Size*. Let us suppose that *Uniformity of Cell Size<=2.5* for the case under investigation. Upon setting this first observation, BayesiaLab will compute the new probability distribution of the **Target Node**, given the evidence. We see that the probability of *Class=Malignant* has increased to 84.81% (Figure 6.54). Given the evidence, BayesiaLab also recomputes the ideal new order of questions and now presents *Bare Nuclei* as the next most relevant question.

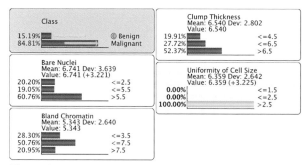

Figure 6.54

Let us now assume that *Bare Nuclei* and *Bland Chromatin* are not available for observation and that we skip answering them. We instead set the node *Clump Thickness<=4.5* (Figure 6.55).

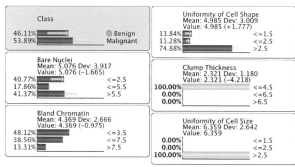

Figure 6.55

Given this latest piece of evidence, the probability distribution of *Class* is once again updated, as is the array of questions. The small gray arrows inside the **Monitors** indicate how the probabilities have changed compared to the prior iteration. (Figure 6.56)

Figure 6.56

It is important to point out that not only the **Target Node** is updated as we set evidence. Rather, all nodes are being updated upon setting evidence, reflecting the omni-directional nature of inference within a Bayesian network. We can continue this process of updating until we have exhausted all available evidence, or until we have reached an acceptable level of certainty regarding the diagnosis.

Chapter 6

WebSimulator

The **Adaptive Questionnaire** was designed as an application that can also be utilized by an "end user" of analysis or study results. As such, it is suitable for a much broader audience, beyond the researcher who uses BayesiaLab as a desktop research laboratory. For instance, clinicians could use the **Adaptive Questionnaire** for decision support without any knowledge of Bayesian networks or information theory.

▶ Model Utilization in Chapter 3, p. 45.

To reach such a broader user group, we can publish the **Adaptive Questionnaire** via BayesiaLab's **WebSimulator** and make it accessible to anyone with an Internet connection.

We continue to use the network developed in this chapter and prepare it for publication. This process is straightforward and only requires declaring what role each node has to play in this use case. We select **Tools > WebSimulator Editor** (Figure 6.57).

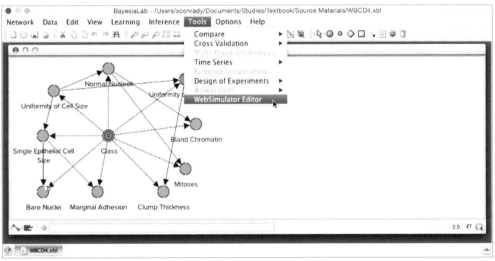

Figure 6.57

In the **WebSimulator Editor**, we need to define **Inputs** and **Outputs**, plus, we need to define the **Questionnaire Targets**. In our example, all predictor nodes serve as **Inputs**. The **Target Node**, *Class*, simultaneously serves as the only **Output** and as the only **Questionnaire Target**. Note that the **Adaptive Questionnaire** in the **WebSimulator** can work with multiple **Targets**.

Additionally, we need to define the type of graphical components that will be used for the **Input** and **Output** nodes. Put simply, the **Inputs** and **Outputs** are web-adaptations of BayesiaLab's **Monitors**, and, as such, there are different ways of setting evidence and reading out posterior probabilities. For the purposes of this example, we select **Discrete States: Dropdown List** for all **Inputs** (Figure 6.58) and **Probabilities: Bar Graph** for the **Output** (Figure 6.59). Most importantly, we mark *Class* as a **Questionnaire Target**.

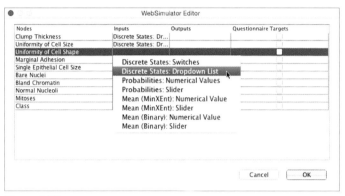

Figure 6.58

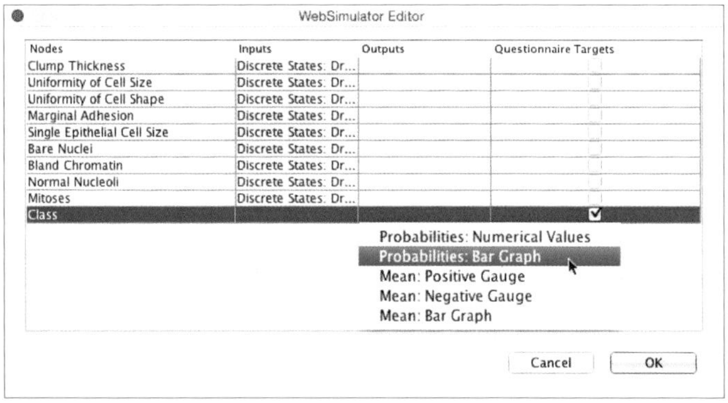

Figure 6.59

This completes the preparation steps in BayesiaLab proper. We save the network—in its native xbl format—to get ready for uploading it to the **BayesiaLab WebSimulator Server**.

We switch to a web browser and open https://simulator.bayesialab.com, which brings up the main **WebSimulator** interface. We click the menu icon and select **Go to Admin** (Figure 6.60), which prompts us to log in (illustration omitted).

Figure 6.60

Chapter 6

Once logged in, we see all models that we have already uploaded to the **WebSimulator** (Figure 6.61).

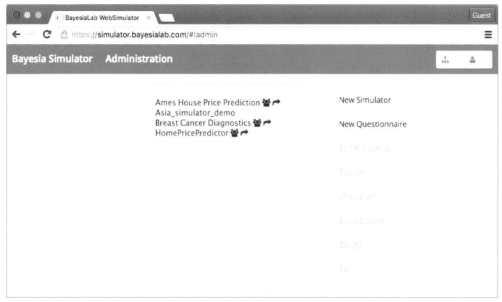

Figure 6.61

Clicking on **New Questionnaire** allows us to select the xbl file we saved and provide names for the questionnaire and the author of the model (Figure 6.62).

Figure 6.62

Upon confirmation, we see our new **Adaptive Questionnaire** listed in the overview of models. From this screen, we can highlight our model and immediately click **Publish**.

153

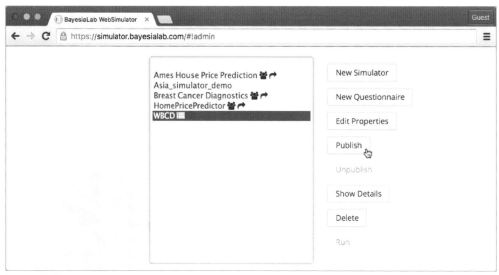

Figure 6.63

As the final step, we need to choose whether our new **Adaptive Questionnaire** should be accessible to the public or whether it needs to be private and password-protected (Figure 6.64). By default, all BayesiaLab users have access to a public account, which is what we select here.

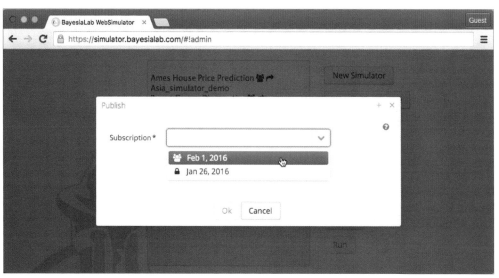

Figure 6.64

Clicking **OK** publishes this questionnaire for the world to see. The questionnaire's public URL can be retrieved by clicking **Show Details** (Figure 6.65).

Chapter 6

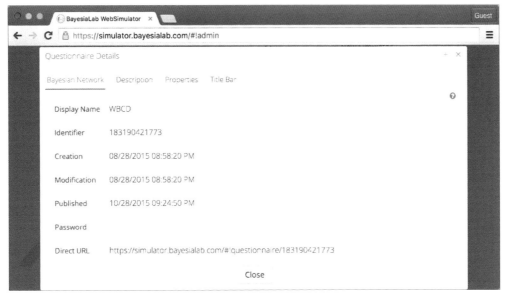

Figure 6.65

The URL of this example,

https://simulator.bayesialab.com/#!questionnaire/183190421773,

can now be used to access the questionnaire from any web browser. The link brings up a simple layout with **Inputs** in the upper left panel and the **Output** in the right-hand panel. Given the input type we chose, we enter our observations via drop-down menus in each **Input**, e.g. *Uniformity of Cell Size>2.5*. This is the same value we used for the earlier demo of the **Adaptive Questionnaire** with the desktop version of BayesiaLab (Figure 6.66).

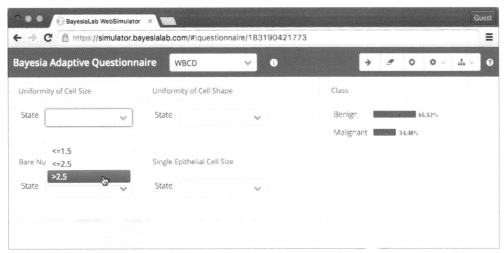

Figure 6.66

155

Upon setting the first piece of evidence, the observed **Input** moves to the bottom left panel, and the probability of the **Output** *Class* is updated. Additionally, the recommended order for the next-best piece of evidence is recomputed and presented in the upper left panel.

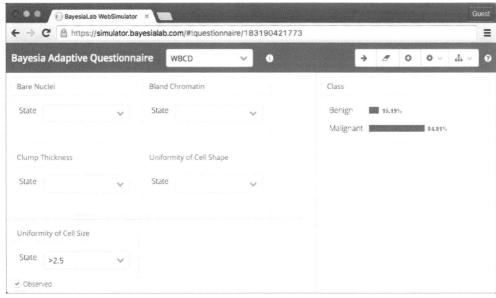

Figure 6.67

Target Interpretation Tree

After this excursion into a web-based application, we return to the BayesiaLab desktop software. However, we remain on the topic of model utilization and show a function that is closely related to the **Adaptive Questionnaire**.

Although its tree structure is not directly visible, the **Adaptive Questionnaire** is a dynamic tree for seeking evidence leading to the diagnosis of one particular case. In fact, a new tree is generated for every extra piece of evidence.

The **Target Interpretation Tree**, on the other hand, is explicitly shown in the form of a static graphical tree. The **Target Interpretation Tree** is induced once from all cases and then prescribes the order for seeking the optimum sequence for gathering evidence. As such, it is not dynamic and not case-specific, i.e. the recommended order does not change given evidence. This makes it practical when hard-and-fast rules are required, e.g. in preparation for emergency situations. However, as a static tree, it lacks the flexibility of skipping missing observations.

Chapter 6

The **Target Interpretation Tree** can be started from the menu via **Analysis > Target Interpretation Tree** (Figure 6.68).

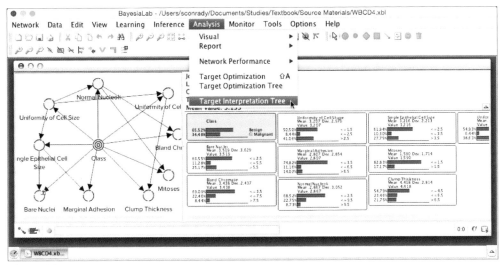

Figure 6.68

Upon starting this function, we need to set several options (Figure 6.69). We define the **Search Stop Criteria** and set the **Maximum Size of Evidence** to 3 and the **Minimum Joint Probability** to 1 (percent).

Figure 6.69

By default, the **Target Interpretation Tree** is presented in a top-down format (Figure 6.70).

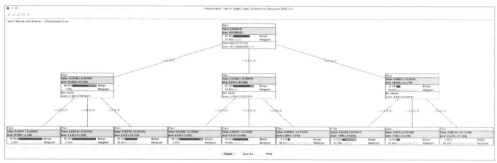

Figure 6.70

Sometimes, it may be more convenient to change the layout to a left-to-right format via the **Switch Position** button ( ) in the upper left-hand corner of the window that contains the tree. The tree in Figure 6.71 is presented in the left-to-right layout.

The **Target Interpretation Tree** prescribes in which sequence evidence should be sought for gaining the maximum amount of information towards a diagnosis, also taking into account the relative cost of acquiring the evidence. Going from left to right, we see how the probability distribution for *Class* changes given the evidence set thus far.

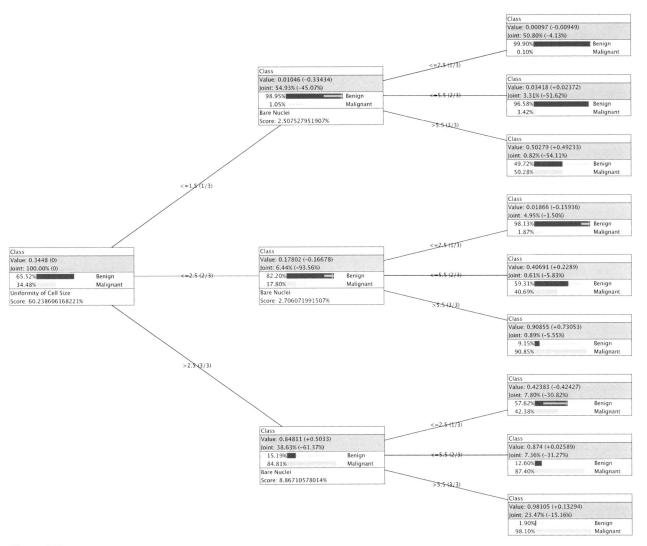

Figure 6.71

The leftmost node in the tree, without any evidence set, shows the marginal probability distribution of *Class*. The bottom panel of this node shows *Uniformity of Cells Size* as the most important evidence to seek.

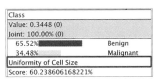

Figure 6.72

The three branches that emerge from the node represent the possible states of *Uniformity of Cells Size*, i.e. the **Hard Evidence** we can observe. If we set evidence analogously to what we did in the **Adaptive Questionnaire**, we will choose the middle branch with the value *Uniformity of Cell Size<=2.5 (2/3)* (Figure 6.73).

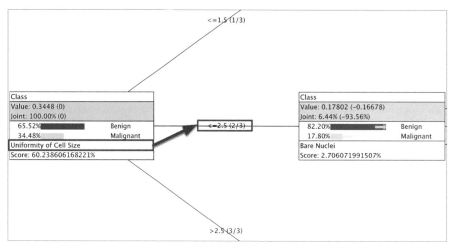

Figure 6.73

This evidence updates the probabilities of the **Target State**, now predicting a 17.80% probability of *Class=Malignant*. At the same time, we can see what is the next best piece of evidence to seek. Here, it is *Bare Nuclei*, which will provide the greatest information gain towards the diagnosis of *Class*. The information gain is quantified with the **Score** displayed at the bottom of the node (Figure 6.74)

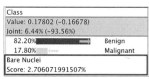

Figure 6.74

The **Score** is the **Conditional Mutual Information** of the node *Bare Nuclei* with regard to the **Target Node**, divided by the cost of observing the evidence if the option **Utilize Evidence Cost** was checked. In our case, as we did not check this option, the **Score** is equal to the **Conditional Mutual Information**.

Mapping

We can quickly verify the Score of 2.7% by running the **Mapping** function. First, we set the evidence *Uniformity of Cell Size<=2.5* and then run **Analysis > Visual > Mapping**.

Chapter 6

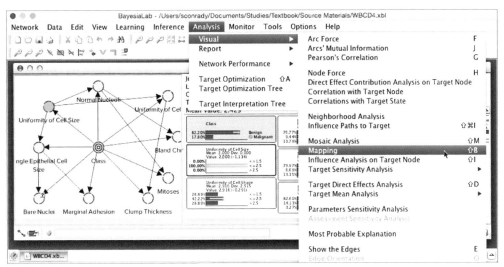

Figure 6.75

The **Mapping** window features drop-down menus for **Node Analysis** and **Arc Analysis** (Figure 6.76). However, we are only interested in **Node Analysis**, and we select **Mutual Information with the Target Node** as the metric to be displayed.

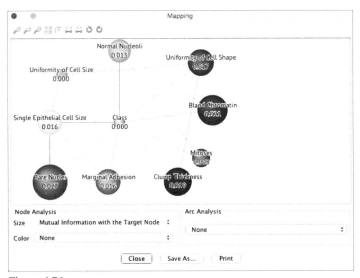

Figure 6.76

The size of the nodes—beyond a fixed minimum size—is now proportional to the **Mutual Information with the Target Node** given the current set of evidence. It shows that given *Uniformity of Cell Size<=2.5*, the **Mutual Information** of *Bare Nuclei* with the **Target Node** is 0.027. Note that the node on which evidence has already been set, i.e. *Uniformity of Cell Size*, shows a **Mutual Information** of 0 (Figure 6.76).

As per this visualization, learning *Bare Nuclei* will bring the highest information gain among the remaining variables. For instance, if we now observed *Bare Nuclei>5.5 (3/3)*, the probability of *Class=Malignant* would reach 90.85% (Figure 6.77).

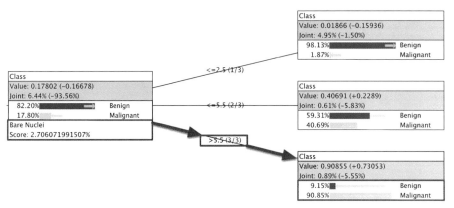

Figure 6.77

Finally, BayesiaLab also reports the joint probability of each tree node, i.e. the probability that all pieces of evidence in a branch, up to and including that tree node, occur (Figure 6.78).

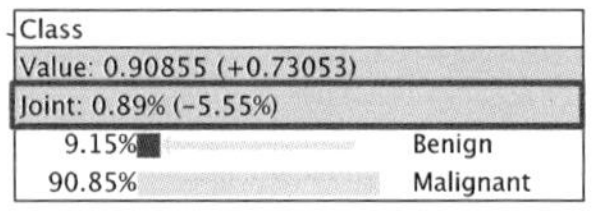

Figure 6.78

Figure 6.78 says that the joint probability of *Uniformity of Cell Size<=2.5* and *Bare Nuclei>5.5* is 0.89%. This implies that in approximately 1 out of 100 case, this particular combination of evidence is to be expected.

As opposed to this somewhat artificial illustration of a **Target Interpretation Tree** in the context of FNA-based diagnosis, **Target Interpretation Trees** are often prepared for emergency situations, such as triage classification, in which rapid diagnosis with constrained resources is essential. We believe that our example still conveys the idea of "optimum escalation" in obtaining evidence towards a diagnosis.

Chapter 6

Chapter 7

7. Unsupervised Learning

Unsupervised Structural Learning is perhaps the purest form of knowledge discovery as there are no hypotheses constraining the exploration of possible relationships between variables. BayesiaLab offers a wide range of algorithms for that purpose. Making this technology easily accessible can potentially transform how researchers approach complex, high-dimensional problem domains.

Example: Stock Market

We find the mass of data available from financial markets to be an ideal proving ground for experimenting with knowledge discovery algorithms that generate Bayesian networks. Comparing machine-learned knowledge with our personal understanding of the stock market can perhaps allow us to validate BayesiaLab's "discoveries." For instance, any structure that is discovered by BayesiaLab's algorithms should be consistent with an equity analyst's understanding of fundamental relationships between stocks.

In this chapter, we will utilize **Unsupervised Learning** algorithms to automatically generate Bayesian networks from daily stock returns recorded over a six-year period. We will examine 459 stocks from the S&P 500 index, for which observations are available over the entire timeframe. We selected the S&P 500 as the basis for our study, as the companies listed on this index are presumably among the best-known corporations worldwide, so even a casual observer should be able to critically review the machine-learned findings. In other words, we are trying to machine learn the obvious, as any mistakes in this process would automatically become self-evident. Quite often, experts' reaction to such machine-learned findings is, "well, we already knew that." Indeed, that is the very point we want to make, as machine-learning can—within seconds—catch up with human expertise accumulated over years and then rapidly expand beyond what is already known.

The power of such algorithmic learning will be still more apparent in entirely unknown domains. However, if we were to machine learn the structure of a foreign equity market for expository purposes, we would probably not be able to judge the resulting structure as plausible or not.

In addition to generating human-readable and interpretable structures, we want to illustrate how we can immediately use machine-learned Bayesian networks as "computable knowledge" for automated inference and prediction. Our objective is to gain both a qualitative and quantitative understanding of the stock market by using Bayesian networks. In the quantitative context, we will also show how BayesiaLab can carry out inference with multiple pieces of uncertain and even conflicting evidence. The ability of Bayesian networks to perform computations under uncertainty makes them suitable for a wide range of real-world applications.

Dataset

The S&P 500 is a free-float capitalization-weighted index of the prices of 500 large-cap common stocks actively traded in the United States, which has been published since 1957. The stocks included in the S&P 500 are those of large publicly held companies that trade on either of the two largest American stock market exchanges; the New York Stock Exchange and the NASDAQ. For our case study, we have tracked the daily closing prices of all stocks included in the S&P 500 index from January 3, 2005, through December 30, 2010, only excluding those stocks that were not traded continuously over the entire study period. This leaves a total of 459 stock prices with 1,510 observations each. The top three panels of Figure 7.1 show the S&P 500 Index, plus the stock prices for Apple Inc. and Yahoo! Inc. Note that the plot of the S&P 500 Index is only shown for reference; the index will not be included in the analysis.

Data Preparation and Transformation

Rather than treating the time series in levels, we difference the stock prices and compute the daily returns. More specifically, we will take differences of the logarithms of the levels, which is a good approximation of the daily stock return in percent. After this transformation, 1,509 observations remain. The bottom three panels of Figure 7.1 display the returns.

Chapter 7

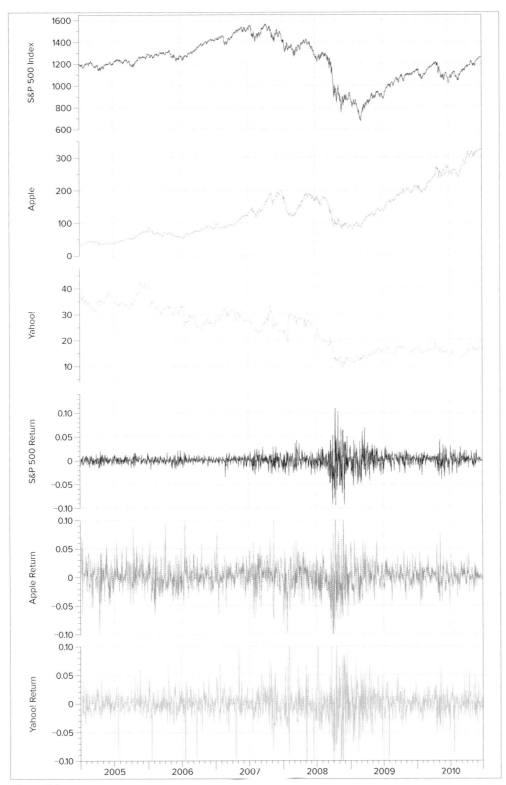

Figure 7.1

Data Import

We use BayesiaLab's **Data Import Wizard** to load all 459 time series[1] into memory from "SP500.csv".[2] BayesiaLab automatically detects the column headers, which contain the ticker symbols[3] as variable names (Figure 7.2).

Figure 7.2

The next step identifies the variable types contained in the database and, as expected, BayesiaLab finds 459 **Continuous** variables (Figure 7.3).

---

1   Although the dataset has a temporal ordering, for expository simplicity, we treat each time interval as an independent observation.

2   The S&P 500 data is available for download from the Bayesia website: www.bayesia.us/sp500

3   A ticker symbol is a short abbreviation used to uniquely identify publicly traded stocks.

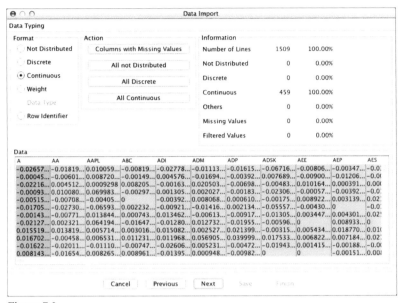

Figure 7.3

There are no missing values in this database, so the next step of the **Data Import Wizard** can be skipped entirely. We still show it in Figure 7.4 for reference, although all options are grayed out.

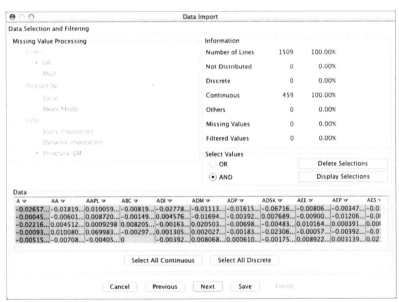

Figure 7.4

Data Discretization

▶ Chapter 9. Missing Values Processing, p. 289.

While we can defer a discussion of **Missing Values Processing** for now, we must carefully consider our options in the next step of the **Data Import Wizard**. Here, we need to discretize all **Continuous** variables, which means all 459 variables in our case. In the context of **Unsupervised Learning**, we do not have a specific target variable. Hence, we have to choose one of the univariate discretization algorithms. Following the recommendations presented in Chapter 5, we choose **K-Means**. Furthermore, given the number of observations that are available, we aim for a discretization with 5 bins, as per the heuristic discussed in Chapter 6.

▶ K-Means in Chapter 5, p. 88.

▶ Discretization Intervals in Chapter 6, p. 119.

While helpful, any such heuristics should not be considered conclusive. Only once a model is learned, we can properly evaluate the adequacy of the selected **Discretization**. In BayesiaLab, we also have access to the discretization functions again anytime after completing the **Data Import Wizard**, which makes experimentation with different discretization methods and intervals very easy (in the **Modeling Mode**, we can start a new discretization with **Learning > Discretization**).

Before proceeding with the automatic discretization of all variables, we shall examine the type of density functions that we can find in this dataset. We use the built-in in plotting function for this purpose, which is available in the next step of the **Data Import Wizard** (Figure 7.5).

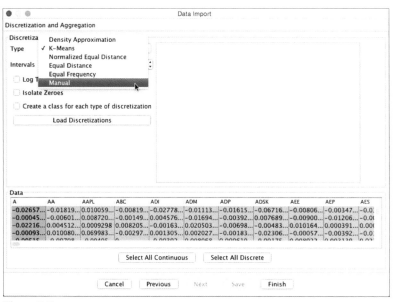

Figure 7.5

Chapter 7

After selecting **Manual** from the **Discretization** drop-down menu (Figure 7.5) and then clicking **Switch View**, we obtain the probability density function of the first variable *A* (Figure 7.6).

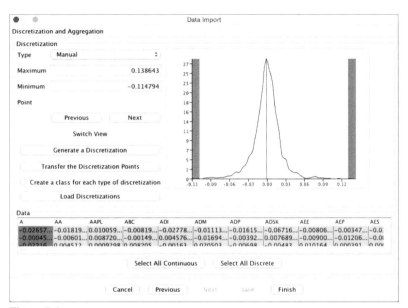

Figure 7.6

Without formally testing it for normality, we judge that the distribution of *A (Agilent Technologies Inc.)* does resemble the shape of a Normal distribution. In fact, the distributions of all variables in this dataset appear fairly similar, which further supports our selection of the **K-Means** algorithm. We click **Finish** to perform the discretization (Figure 7.7). A progress bar is shown to report on the progress.

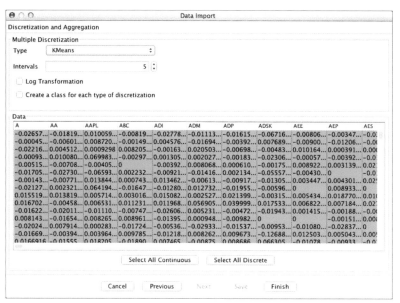

Figure 7.7

Modeling Mode

After completing the data import, the variables are delivered into the **Graph Panel** (Figure 7.8). The original variable names, which were stored the first line of the database, become our **Node Names**.

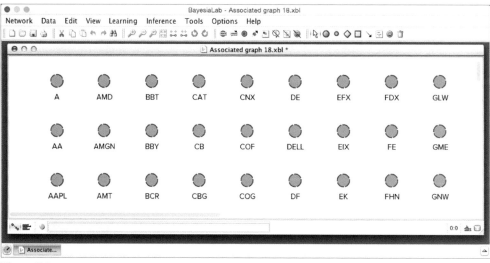

Figure 7.8

▸ Node Comments in Chapter 5, p. 92.

At this point, it is practical to add **Node Comments** to associate full company names with the ticker symbols. We use a Dictionary file for that purpose (Figure 7.9).

Chapter 7

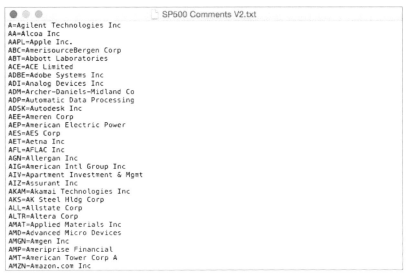

Figure 7.9

This file can be loaded into BayesiaLab via **Data > Associate Dictionary > Node > Comments** (Figure 7.10).

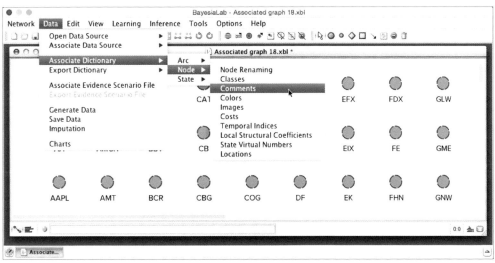

Figure 7.10

Once the **Node Comments** are loaded, a small call-out symbol (ⓘ) appears next to each **Node Name**, confirming that associating the **Dictionary** completed successfully. (Figure 7.11).

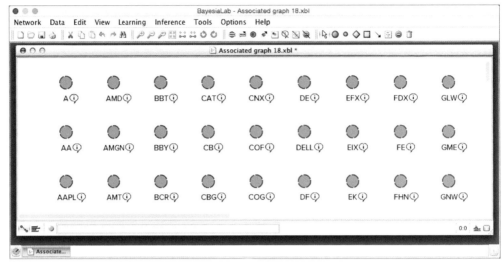

Figure 7.11

As the name implies, selecting **View > Display Node Comments** (⊕ or [M]) reveals the full company names (Figure 7.12).

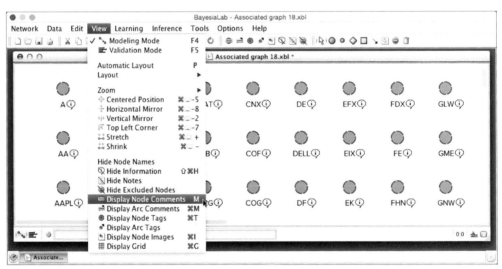

Figure 7.12

Node Comments can be displayed for either all nodes or only for selected ones (Figure 7.13).

Chapter 7

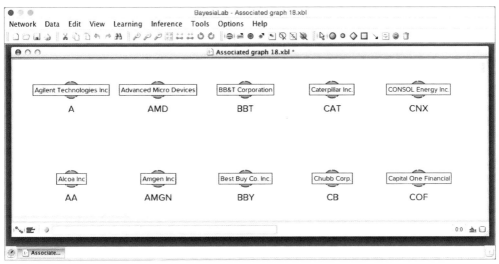

Figure 7.13

Data Import Review

Before proceeding with the first learning step, we recommend switching to the **Validation Mode** (or) to verify the results of the import and discretization (Figure 7.14).

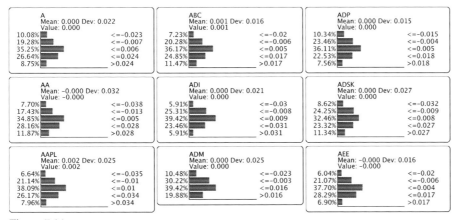

Figure 7.14

This gives us an opportunity to compare the variables' statistics with our understanding of the domain. At first glance, mean values of near zero for all distributions might suggest that stocks prices remained "flat" throughout the observation period. For the S&P 500 index, this was actually true. However, it could not be true for all individual stocks given that the Apple stock, for instance, increased ten-fold in value between 2005 and 2010. The seemingly low returns are due to the fact that we are studying

175

daily returns, rather than annual returns. On that basis, even the rapid appreciation of the Apple stock translates into an average daily return of "only" 0.2%.[4]

A "sanity check" of this kind is the prudent thing to do before proceeding to machine-learning.

Unsupervised Learning

The computational complexity of BayesiaLab's **Unsupervised Learning** algorithms exhibits quadratic growth as a function of the number of nodes. However, the **Maximum Weight Spanning Tree (MWST)** is constrained to learning a tree structure (one parent per node), which makes it much faster that the other algorithms. More specifically, the **MWST** algorithm includes only one procedure with quadratic complexity, namely the initialization procedure that computes the matrix of bivariate relationships.

Given the number of variables in this dataset, we decide to use the **MWST**. Performing the **MWST** algorithm with a file of this size should only take a few seconds. Moreover, using BayesiaLab's layout algorithms, the tree structures produced by **MWST** can be easily transformed into easy-to-interpret layouts. Thus, **MWST** is a practical first step for knowledge discovery. Furthermore, this approach can be useful for verifying that there are no coding problems, e.g. with variables that are unconnected. Given the quick insights that can be gleaned from it, we recommend using **MWST** at the beginning of most studies.

We switch back to **Modeling Mode** (✎ or [F4]) and select **Learning > Unsupervised Structural Learning > Maximum Spanning Tree**.

---

4   BayesiaLab reports the arithmetic mean as **Value** in the **Monitors**.

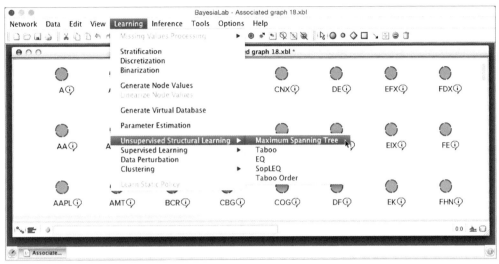

Figure 7.15

In addition to its one-parent constraint, **MWST** is also unique in that it is the only learning algorithm in BayesiaLab that allows us to choose the scoring method for learning, i.e. **Minimum Description Length** (MDL) or **Pearson's Correlation** (Figure 7.16). Unless we are certain about the linearity of the yet-to-be-learned relationships between variables, **Minimum Description Length** is the better choice and, hence, the default setting.

Figure 7.16

At first glance, the resulting network does not appear simple and tree-like at all (Figure 7.17).

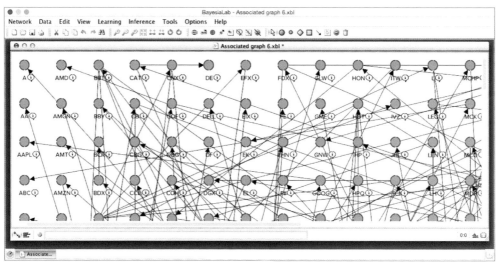

Figure 7.17

This can be addressed with BayesiaLab's built-in layout algorithms. Selecting **View > Automatic Layout** (P) quickly rearranges the network to reveal the tree structure. The resulting reformatted Bayesian network can now be readily interpreted (Figure 7.18).

Chapter 7

Figure 7.18

Network Analysis

Let us suppose we are interested in *Procter & Gamble (PG)*. First, we look for the corresponding node using the **Search** function (`Control` `F`). Note that we will be able to search for the full company name if we check **Include Comments**. Furthermore, we can use a combination of wildcards in the search, e.g. "*" as a placeholder for a character string of any length or "?" for a single character (Figure 7.19).

Selecting *PG* from the listing search results makes the corresponding node flash for a few seconds so it can be found among the hundreds of nodes on the screen.

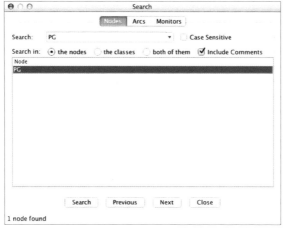

Figure 7.19

Once located, we can zoom in to see *PG* and numerous adjacent nodes.

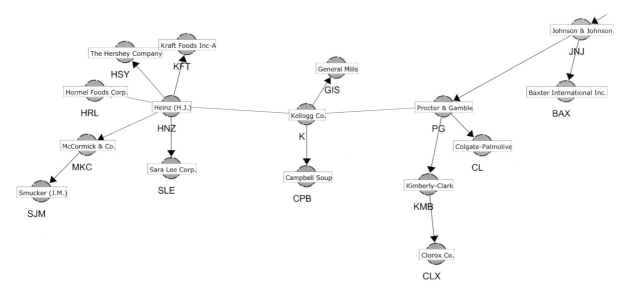

Figure 7.20

As it turns out, the "neighborhood" of *Procter & Gamble* contains many familiar company names, mostly from the consumer packaged goods industry. Perhaps these companies appear all-too-obvious, and one might wonder what insight we gained at this point. The chances are that even a casual observer of the industry would have mentioned *Kimberly-Clark*, *Colgate-Palmolive*, and *Johnson & Johnson* as businesses operating in the same field as *Procter & Gamble*. Therefore, one might argue, similar stock price movements should be expected.

The key point here is that—without any prior knowledge of this domain—a computer algorithm automatically extracted a structure that is consistent with the understanding of anyone familiar with this domain.

Beyond interpreting the qualitative structure of this network, there is a wide range of functions for gaining insight into this high-dimensional problem domain. For instance, we may wish to know which node within this network is most important. In Chapter 5, we discussed the question in the context of a predictive model, which we learned with **Supervised Learning**. Here, on the other hand, we learned the network with an **Unsupervised Learning** algorithm, which means that there is no **Target Node**. As a result, we need to think about the importance of a node with regard to the entire network, as opposed to a specific **Target Node**.

We need to introduce a number of new concepts to equip us for the discussion about node importance within a network. As we did in Chapter 5, we once again draw on concepts from information theory.

Arc Force

BayesiaLab's **Arc Force** is computed by using the **Kullback-Leibler Divergence**, denoted by D_{KL}, which compares two joint probability distributions, P and Q, defined on the same set of variables $\mathcal{X}$.

$$D_{KL}(P(\mathcal{X}) \| Q(\mathcal{X})) = \sum_{\mathcal{X}} P(\mathcal{X}) \log_2 \frac{P(\mathcal{X})}{Q(\mathcal{X})}, \tag{7.1}$$

where P is the current network, and Q is the exact same network as P, except that we removed the arc under study.

It is important to point out that **Mutual Information** and **Arc Force** are closely related. If the child node in the pair of nodes under study does not have any other parents, **Mutual Information** and **Arc Force** are, in fact, equivalent. However, the **Arc Force** is more powerful as a measure as it takes into account the network's joint probability distribution, rather than only focusing on the bivariate relationship.

The **Arc Force** can be displayed directly on the Bayesian network graph. Upon switching to the **Validation Mode** (☰ or [F5]), we select **Analysis > Visual > Arc Force** ([F]) (Figure 7.21).

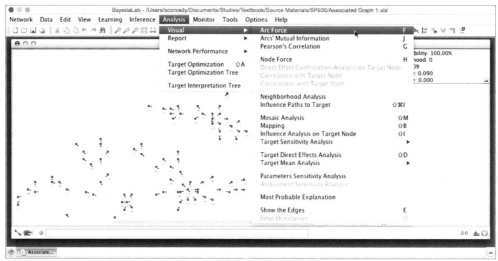

Figure 7.21

Upon activating **Arc Force**, we can see that the arcs have different thicknesses. Also, an additional control panel becomes available in the menu (Figure 7.22).

Figure 7.22

The slider in this control panel allows us to set the **Arc Force** threshold below which arcs and nodes will be grayed out in the **Graph Panel**. By default, it is set to 0, which means that the entire network is visible. Using **Previous** (◁) and **Next** (▷), we can step through all threshold levels. For instance, by starting at the maximum and then going down one step, we highlight the arc with the strongest **Arc Force** in the this network, which is between *SPG (Simon Property Group)* and *VNO (Vornado Realty Trust)* (Figure 7.23).

Chapter 7

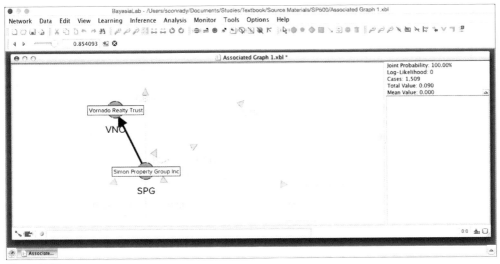

Figure 7.23

Node Force

The **Node Force** can be derived directly from the **Arc Force**. More specifically, there are three types of **Node Force** in BayesiaLab (the corresponding menu icons are shown in parentheses):

- The **Incoming Node Force** ( ) is the sum of the **Arc Forces** of all incoming arcs.
- The **Outgoing Node Force** ( ) is the sum of the **Arc Forces** of all outgoing arcs.
- The **Total Node Force** is the ( ) sum of the **Arc Forces** of all incoming and outgoing arcs.

The **Node Force** can be shown directly on the Bayesian network graph. Upon switching to the **Validation Mode** (or [F5]), we select **Analysis > Visual > Node Force** ([H]).

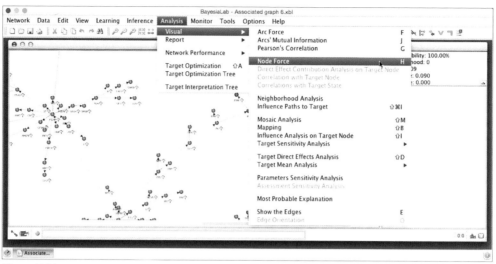

Figure 7.24

After starting **Node Force**, we have another additional control panel available in the menu (Figure 7.25).

Figure 7.25

The slider in this control panel allows us to set the *Node Force* threshold below which nodes will be grayed out in the **Graph Panel**. By default, it is set to 0, which means that all nodes are visible. Conversely, by setting the threshold to the maximum, all nodes are grayed out. Using **Previous** (◁) and **Next** (▷), we can step through the entire range of thresholds. For example, by starting at the maximum and then going down one step, we can find the node with the strongest **Node Force** in the this network, which is *BEN (Franklin Resources)*, a global investment management organization (Figure 7.26). This functionality is analogous to the control panel for **Arc Force**.

Chapter 7

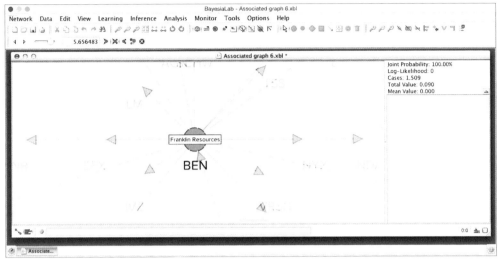

Figure 7.26

Node Force Mapping

This analysis tool also features a "local" **Mapping** function, which is particularly useful when dealing with big networks, such as the one in this example with hundreds of nodes. We refer to this as a "local" **Mapping** function in the sense of only being available in the context of **Node Force Analysis**, as opposed to the "general" **Mapping** function, which is always available within the **Validation Mode** as a standalone analysis tool (**Analysis > Visual > Mapping**).

▸ Mapping in Chapter 6, p. 160.

We launch the **Mapping** window by clicking the **Mapping** icon ( ) on the control panel, to the right of the slider. In this network view, the size of the nodes is directly proportional to the selected type of **Node Force** (Incoming, Outgoing, Total). The thickness of the links is proportional to the **Arc Force**. Changing the threshold values (with the slider for example) automatically updates the view.

Choosing **Static Font Size** from the **Contextual Menu** and then, for instance, reducing the threshold by four more steps, reveals the five strongest nodes, while maintaining an overview of the entire network (Figure 7.27).

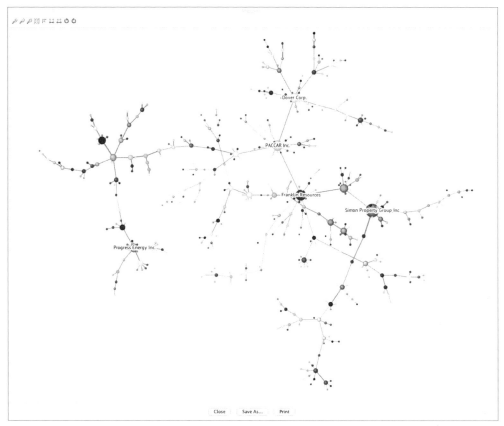

Figure 7.27

Inference

Throughout this book so far, we have performed inference with various types of evidence. The kind of evidence we used was routinely determined by the nature of the problem domain.

We will now use this example to systematically try out all types of evidence for performing inference. With that, we depart from our habit of showing only realistic applications. One could certainly argue that not all types of evidence are plausible in the context of a Bayesian network that represents the stock market. In particular, any inference we perform here with arbitrary evidence should not be interpreted as an attempt to predict stock prices. Nevertheless, for the sake of an exhaustive presentation, even this somewhat contrived exercise shall be educational.

Chapter 7

Within our large network of 459 nodes, we will only focus on a small subset of nodes, namely *PG (Procter & Gamble)*, *JNJ (Johnson & Johnson)*, and *KMB (Kimberly-Clark)*. These nodes come from the "neighborhood" shown in Figure 7.20.

We start by highlighting *PG*, *JNJ*, and *KMB* to bring up their **Monitors**. Prior to setting any evidence, we see their marginal distributions in the **Monitors**. We see that the expected value (mean value) of the returns is 0 (Figure 7.28).

▸ Data Import Review, p. 175.

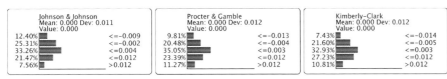

Figure 7.28

Inference with Hard Evidence

Next, we double-click the state *JNJ>0.012* to compute the posterior probabilities of *PG* and *KMB* given this evidence. The gray arrows indicate how the distributions have changed compared to before, prior to setting evidence (Figure 7.29). Given the evidence, the expected value of *PG* and KMB are 1.2% and 0.6% respectively (Figure 7.29).

Figure 7.29

If we also set *KMB* to its highest state (*KMB>0.012*), this would further reduce the uncertainty of *PG* and compute an expected value of 1.8% (Figure 7.30). This means that *PG* had an average daily return of 1.8% on days when this evidence was observed.

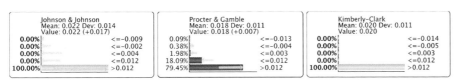

Figure 7.30

Inference with Probabilistic and Numerical Evidence

Given the discrete states of nodes, setting **Hard Evidence** is presumably intuitive to understand. However, the nature of many real-world observations calls for so-called **Probabilistic** or **Numerical Evidence**. For instance, the observations we make in a domain can include uncertainty. Also, evidence scenarios can consist of values that do not coincide with the values of nodes' states. So, as an alternative to **Hard Evidence**, we can use BayesiaLab to set such evidence.

▶ Types of Evidence in Chapter 3, p. 42.

Probabilistic Evidence

Probabilistic Evidence is a convenient way for directly encoding our assumptions about possible conditions of a domain. For example, a stock market analysts may consider a scenario with a specific probability distribution for *JNJ* corresponding to a hypothetical period of time (i.e. a subset of days). Given his understanding of the domain, he can assign probabilities to each state, thus encoding his belief.

After removing the prior evidence ( ), we can set such beliefs as **Probabilistic Evidence** by right-clicking the *JNJ* **Monitor** and then selecting **Enter Probabilities** (Figure 7.31).

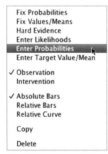

Figure 7.31

For the distribution of **Probabilistic Evidence**, the sum of the probabilities must be equal to 100%. We can adjust the **Monitor's** bar chart by dragging the bars to the probability levels that reflect the scenario under consideration (Figure 7.31). By double-clicking on the percentages, we can also directly enter the desired probabilities. Note that changing the probability of any state automatically updates the probabilities of all other states to maintain the sum constraint.

Figure 7.32

In order to remove a degree of freedom in the sum constraint, we left-click the **State Name/Value** in the **Monitor**, to the right of each bar (Figure 7.33). Doing so locks the currently set probability and turns the corresponding bar turns green. The probability of this state will no longer be automatically updated while the probabilities of other states are being edited. This feature is essential for defining a distribution on nodes that have more than two states. Another left click on the same **State Name/Value** unlocks the probability again.

Figure 7.33

There are two ways to validate the entered distribution, via the green and the purple buttons. Clicking the green button (☐) defines a static likelihood distribution. This means that any additional piece of evidence on other nodes can update the distribution we set (Figure 7.34).

Figure 7.34

Clicking the purple button (☐) "fixes" the probability distribution we entered by defining dynamic likelihoods. This means that each new piece of evidence triggers an update of the likelihood distribution in order to maintain the same probability distribution (Figure 7.35).

Figure 7.35

Using either validation method, BayesiaLab computes a likelihood distribution that produces the requested probability distribution. By setting this distribution, BayesiaLab also performs inference automatically and updated the probabilities of the other nodes in the network (Figure 7.36).

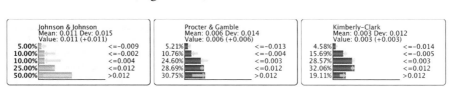

Figure 7.36

Numerical Evidence

Instead of a specific probability distribution, an observation or scenario may exist in the form of a single numerical value, which mans that we need to set **Numerical Evidence**. For instance, a stock market analyst may wish to examine how other stocks performed given a hypothetic period of time during which the average of the daily returns of *JNJ* was −1%. Naturally, this requires that we set evidence on *JNJ* that has an expected (mean) value of −0.01 (=−1%). However, this task is not as straightforward as it may sound. The questions will become apparent as we go through the steps to set this evidence.

First, we right-click *JNJ* **Monitor** and then select Enter **Target Value/Mean** from the **Contextual Menu** (Figure 7.37).

Figure 7.37

Next, we type "−0.01" into the dialog box for **Target Mean/Value** (Figure 7.38). Additionally, as was the case with **Probabilistic Evidence**, we have to choose the type of validation, but we now have three options under **Observation Type**:

- **No Fixing**, which is the same as the green button, i.e. validation with static likelihood.
- **Fix Mean**, which is the same as the purple button, except that the likelihood is dynamically computed to maintain the mean value, although the probability distribution can change as a result of setting additional evidence.
- **Fix Probabilities**, which is the same as the purple button, i.e. validation with dynamic likelihood.

Figure 7.38

Apart from setting the validation method, we also need to choose the **Distribution Estimation Method** as we need to come up with a distribution that produces the desired mean value. Needless to say, there is a great number of distributions that could potentially produce a mean value of −0.01. However, which one is appropriate?

To make a prudent choice, we first need to understand what the evidence represents. Only then can we choose from the three available algorithms for generating the **Target Distribution** that will produce the **Target Mean/Value**.

MinXEnt ("Minimum Cross-Entropy")

Using the **MinXEnt** algorithm, the **Target Distribution**, which produces the **Target Mean/Value**, is computed in such a way that the **Cross-Entropy** between the original probability distribution of the node and the **Target Distribution** is minimized. Figure 7.39 shows the distribution with a mean of −0.01 that is "closest" in terms of **Cross-Entropy** to the original, marginal distribution shown earlier in Figure 7.28.

Figure 7.39

Binary

The **Target Mean/Value** is generated by interpolating between values of two adjacent states, hence the name "Binary." Here, a "mix" of the values of two states, i.e. *JNJ<=–0.009* and *JNJ<=–0.002,* produces the desired mean of –0.01.

Figure 7.40

Value Shift

The **Target Mean/Value** is generated by shifting the values of each particle (or virtual observation) by the exact same amount.

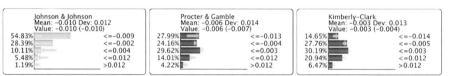

Figure 7.41

Target Value/Mean Considerations

As we see in the examples above, using different **Target Distributions** as **Numerical Evidence**—albeit with the same mean value—results in different probability distributions.

Binary

The **Binary** algorithm produces the desired value through interpolation, as in Fuzzy Logic. Among the three available methods, it generates distributions that have the lowest degree of uncertainty. Using the **Binary** algorithm for generating a **Target Mean/Value** would be appropriate if two conditions are met:

1. There is no uncertainty regarding the evidence, i.e. we want the evidence to represent a specific numerical value. "No uncertainty" would typically

apply in situations in which we want to simulate the effects of nodes that represent variables under our control.

2. The desired numerical value is not directly available by setting **Hard Evidence**. In fact, a distribution produced by the **Binary** algorithm would coincide with **Hard Evidence** if the requested **Target Value/Mean** precisely matched the value of a particular state.

Given that is impossible to directly set prices in the stock market, it is clearly not a good example for illustrating this as a use of the **Binary** algorithm. Perhaps a price elasticity model would be more appropriate. In such a model, we would want to infer sales volume based on one specific price level as opposed to a broad range of price levels within a distribution.

MinXEnt and Value Shift

The other two algorithms, **MinXEnt** and **Value Shift**, generate **Soft Evidence**. This means that the **Target Distribution** they supply should be understood like posterior distribution given evidence set on a "hidden cause", i.e. evidence on a variable not included in the model. As such, using **MinXEnt** or **Value Shift** is suitable for creating evidence that represents changing levels of measures like customer satisfaction. Unlike setting the price of a product, we cannot directly adjust the satisfaction of all customers to a specific level. This would imply setting an unrealistic distribution with low or no uncertainty.

More realistically, we would have to assume that higher satisfaction is the result of an enhanced product or better service, i.e. a cause from outside the model. Thus, we need to generate the evidence for customer satisfaction as if it were produced by a hidden cause. This also means that **MinXEnt** and **Value Shift** will produce a distribution close to the marginal one if the targeted **Numerical Evidence** is close to the marginal value.

Special Cases of Numerical Evidence

If the **Numerical Evidence** is equal to current expected value, using (a) **MinXEnt** or (b) **Value Shift** will not change the distribution (Figure 7.42). Using the **Binary** algorithm (c), however, will return a different distribution (except in the context of a binary node).

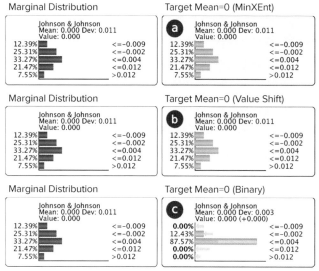

Figure 7.42

Conflicting Evidence

In the examples shown so far, setting evidence typically reduced uncertainty with regard to the node of interest. Just by visually inspecting the distributions, we can tell that setting evidence generally produces "narrower" posterior probabilities.

However, this is not always the case. Occasionally, separate pieces of evidence can conflict with each other. We illustrate this by setting such evidence on *JNJ* and *KMB*. We start with the marginal distribution of all nodes (Figure 7.43).

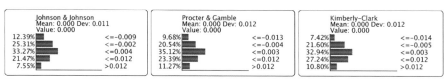

Figure 7.43

After setting **Numerical Evidence** (using **MinXEnt**) with a **Target Mean/Value** of +1.5% on *JNJ*, we obtain Figure 7.44.

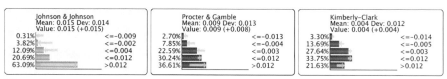

Figure 7.44

The posterior probabilities inferred as a result of the *JNJ* evidence indicate that the *PG* distribution is more positive than before. More importantly, the uncertainty regarding *PG* is lower. A stock market analyst would perhaps interpret the *JNJ* movement as a positive signal and hypothesize about a positive trend in the *CPG* industry. In an effort to confirm his hypothesis, he would probably look for additional signals that either confirm the trend and the related expectations regarding *PG* and similar companies.

In the *KMB* **Monitor**, the gray arrows and "(+0.004)" indicate that the first evidence increases the expectation that *KMB* will also increase in value. If we observed, however, that *KMB* decreased by 1.5% (once again using **MinXEnt**), this would go against our expectation (Figure 7.45).

Figure 7.45

The result is that we now have a more uniform probability distribution for *PG*—rather than a narrower distribution. This increases our uncertainty about the state of *PG* compared to the marginal distribution (Figure 7.43).

Even though it appears that we have "lost" information by setting these two pieces of evidence, we may have a knowledge gain after all: we can interpret the uncertainty regarding *PG* as a higher expectation of volatility.

Measures of Conflict

Beyond a qualitative interpretation of contradictory evidence, our Bayesian network model allows us to examine "conflict" beyond its common-sense meaning. A formal conflict measure can be defined by comparing the joint probabilities of the current model versus a reference model, given the same set of evidence for both.

A fully unconnected network is commonly used as the reference model, the so-called "straw model." It is a model that considers all nodes to be marginally independent. If the joint probability of the set of evidence returned by the model under study is lower than that of the reference model, we determine that we have a conflict. Otherwise, if the joint probability is higher, we conclude that the pieces of evidence are consistent.

The conflict measures that are available in BayesiaLab are formally defined as follows:

Global Conflict (Overall Conflict)

$$GC(E) = \log_2 \frac{\prod_{i=1}^{n} P(e_i \mid e_{i-1},...,e_1)}{\prod_{i=1}^{n} P(e_i)}, \tag{7.2}$$

where E is the current set of evidence consisting of n observations, and e_i is the ith piece of evidence.

Bayes Factor

$$BF(E,h) = \log_2 \frac{P(h \mid E)}{P(h)}, \tag{7.3}$$

where h is a hypothetical piece of evidence that has not yet been set or observed.

Local Conflict (Local Consistency)

$$LC(E,h) = GC(E) + BF(E,h) \tag{7.4}$$

$$= \log_2 \frac{\prod_{i=1}^{n} P(e_i \mid e_{i-1},...,e_1) P(h \mid E)}{\prod_{i=1}^{n} P(e_i) P(h)}$$

Evidence Analysis Report

Using these definitions, we can compute to what extent a new observation would be consistent with the current set of evidence. BayesiaLab provides us with this capability in the form of the **Evidence Analysis Report**, which can be generated by selecting **Analysis > Report > Evidence Analysis** from the main menu (Figure 7.46).

Chapter 7

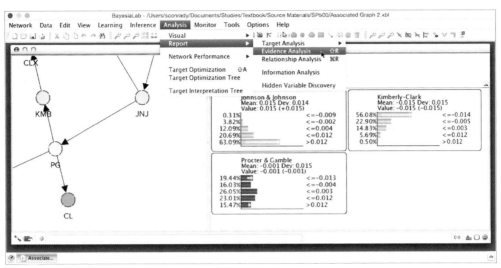

Figure 7.46

The **Evidence Analysis Report** displays two closely related metrics, **Local Consistency (LC)** and the **Bayes Factor (BF)**, for each state of each unobserved node in the network, given the set of evidence. The top portion of this report is shown in Figure 7.47. Also, as we anticipated, an **Overall Conflict** between the two pieces of evidence is shown at the top of the report.

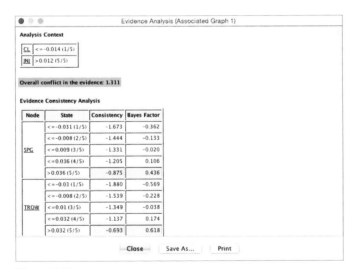

Figure 7.47

Summary

Unsupervised Learning is a practical approach for obtaining a general understanding of simultaneous relationships between many variables in a database. The learned

Bayesian network facilitates visual interpretation plus computation of omni-directional inference, which can be based on any type of evidence, including uncertain and conflicting observations. Given these properties, **Unsupervised Learning** with Bayesian networks becomes a universal tool for knowledge discovery in high-dimensional domains.

Chapter 7

Chapter 8

8. Probabilistic Structural Equation Models

Structural Equation Modeling is a statistical technique for testing and estimating causal relations using a combination of statistical data and qualitative causal assumptions. This definition of a Structural Equation Model (SEM) was articulated by the geneticist Sewall Wright (1921), the economist Trygve Haavelmo (1943) and the cognitive scientist Herbert Simon (1953), and formally defined by Judea Pearl (2000). Structural Equation Models (SEM) allow both confirmatory and exploratory modeling, meaning they are suited to both theory testing and theory development.

What we call Probabilistic Structural Equation Models (PSEMs) in BayesiaLab are conceptually similar to traditional SEMs. However, PSEMs are based on a Bayesian network structure as opposed to a series of equations. More specifically, PSEMs can be distinguished from SEMs in terms of key characteristics:

- All relationships in a PSEM are probabilistic—hence the name, as opposed to having deterministic relationships plus error terms in traditional SEMs.
- PSEMs are nonparametric, which facilitates the representation of nonlinear relationships, plus relationships between categorical variables.
- The structure of PSEMs is partially or fully machine-learned from data.

In general, specifying and estimating a traditional SEM requires a high degree of statistical expertise. Additionally, the multitude of manual steps involved can make the entire SEM workflow extremely time-consuming. The PSEM workflow in BayesiaLab, on the other hand, is accessible to non-statistician subject matter experts. Perhaps more importantly, it can be faster by several orders of magnitude. Finally, once a PSEM is validated, it can be utilized like any other Bayesian network. This means that the full array of analysis, simulation, and optimization tools is available to leverage the knowledge represented in the PSEM.

Example: Consumer Survey

In this chapter, we present a prototypical PSEM application: key drivers analysis and product optimization based on consumer survey data. We examine how consumers perceive product attributes, and how these perceptions relate to the consumers' purchase intent for specific products.

Given the inherent uncertainty of survey data, we also wish to identify higher-level variables, i.e. "latent" variables that represent concepts, which *are not* directly measured in the survey. We do so by analyzing the relationships between the so-called "manifest" variables, i.e. variables that *are* directly measured in the survey. Including such concepts helps in building more stable and reliable models than what would be possible using manifest variables only.

Our overall objective is making surveys clearer to interpret by researchers and making them "actionable" for managerial decision makers. The ultimate goal is to use the generated PSEM for prioritizing marketing and product initiatives to maximize purchase intent.

Dataset

This study is based on a monadic[1] consumer survey about perfumes, which was conducted by a market research agency in France. In this example, we use survey responses from 1,320 women who have evaluated a total of 11 fragrances (representative of the French market) on a wide range of attributes:

- 27 ratings on fragrance-related attributes, such as, *Sweet, Flowery, Feminine*, etc., measured on a 1–10 scale.
- 12 ratings with regard to imagery about someone who wears the respective fragrance, e.g. *Sexy, Modern*, measured on a 1–10 scale.
- 1 variable for *Intensity*, a measure reflecting the level of intensity, measured on a 1–5 scale.[2]
- 1 variable for *Purchase Intent*, measured on a 1–6 scale.
- 1 nominal variable, *Product*, for product identification.

Workflow Overview

A PSEM is a hierarchical Bayesian network that can be generated through a series of machine-learning and analysis tasks:

---

1 A product test only involving one product. In this study, each respondent evaluated only one perfume.

2 The variable *Intensity* is listed separately due to the a-priori knowledge of its non-linearity and the existence of a "just-about-right" level.

Chapter 8

1. **Unsupervised Learning**, to discover the strongest relationships between the manifest variables.
2. **Variable Clustering**, based on the learned Bayesian network, to identify groups of variables that are strongly connected.
3. **Multiple Clustering**: we consider the strong intra-cluster connections identified in the **Variable Clustering** step to be due to a "hidden common cause." For each cluster of variables, we use **Data Clustering**—on the variables within the cluster only—to induce a latent variable that represents the hidden cause.
4. **Unsupervised Learning**, to find the interrelations between the newly-created latent variables and their relationships with the **Target Node**.

Data Import

We have already described all steps of the **Data Import Wizard** in previous chapters. Therefore, we present most of the following screenshots without commentary and only highlight items that are specific to this example. To start the **Data Import Wizard**, we open the file "perfume.csv."[3]

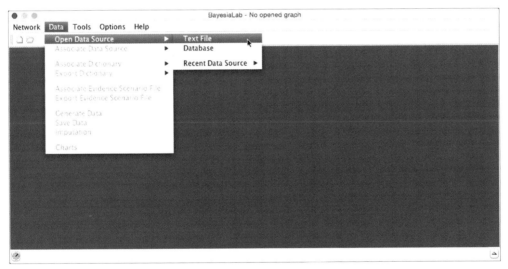

Figure 8.1

---

3 The perfume study data is available for download from the Bayesia website: www.bayesia.us/perfume

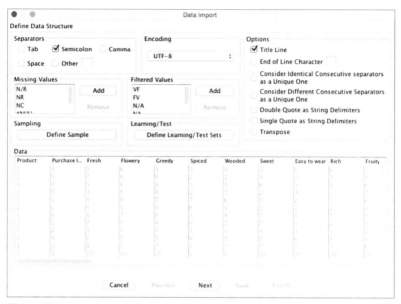

Figure 8.2

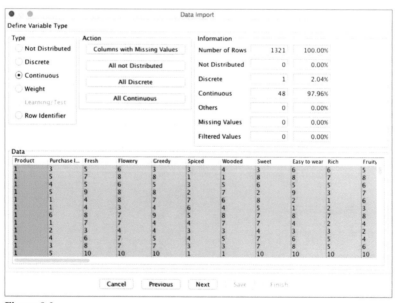

Figure 8.3

For this example, we need to override the default data type for the variable named *Product* as it is a nominal product identifier rather than a numerical value. We can change this variable's data type by highlighting the *Product* column and clicking the **Discrete** radio button. This changes the color of the *Product* column to red (Figure

8.4). We also define *Purchase Intent* and *Intensity* as **Discrete** variables. Their number of states is suitable for our purposes.

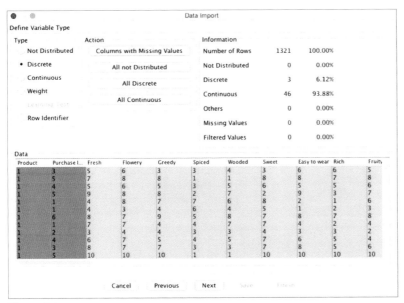

Figure 8.4

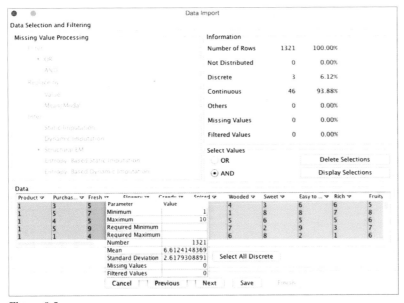

Figure 8.5

The next step is the **Discretization and Aggregation** screen. Given the number of observations, it is appropriate to reduce the number of states of the ratings from the original 10 states (1–10) to a smaller number. All these variables are measuring satis-

▶ Equal Distance in Chapter 5, p. 87.

faction on the same scale, i.e. from 1 to 10. Following our earlier recommendations, the best choice in this context is the **Equal Distance** discretization.

By clicking **Select All Continuous**, we highlight all to-be-discretized variables. Then, we choose the type of discretization to be applied, which is **Equal Distance** (Figure 8.6). Furthermore, given the number of observations, we choose 5 bins for the discretization.

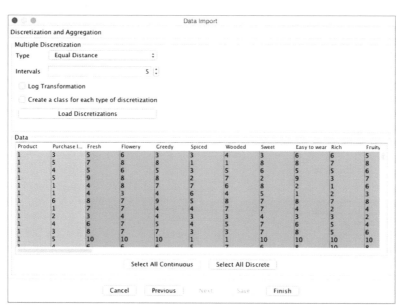

Figure 8.6

Clicking **Finish** finalizes the import process. Upon completion, we are prompted whether we want to view the **Import Report** (Figure 8.7)

Figure 8.7

As there is no uncertainty with regard to the outcome of the discretization, we decline and automatically obtain a fully unconnected network with 49 nodes (Figure 8.8).

Chapter 8

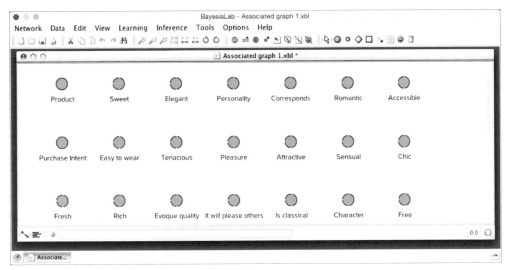

Figure 8.8

Step 1: Unsupervised Learning

As a first step, we need to exclude the node *Purchase Intent*, which will later serve as our **Target Node**. We do not want this node to become part of the structure that we will subsequently use for discovering hidden concepts. Likewise, we need to exclude the node *Product*, as it does not contain consumer feedback to be evaluated.

We can exclude nodes by selecting the **Node Exclusion Mode** (⊖) and then clicking on the to-be-excluded node (Figure 8.9). Alternatively, holding [X] while clicking the node performs the same function.

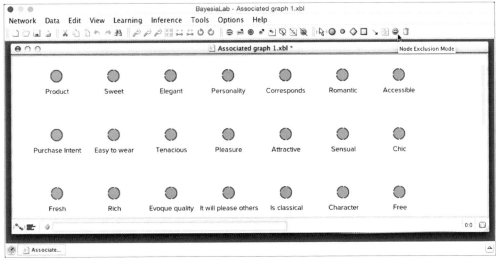

Figure 8.9

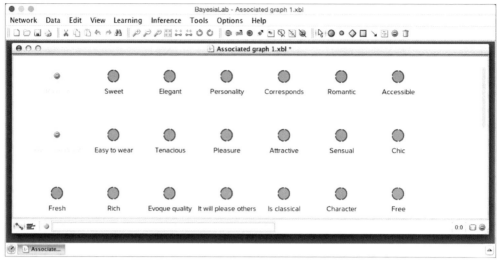

Figure 8.10

The upcoming series of steps is crucial. We now need to prepare a robust network on which we can later perform the clustering process. Given the importance, we recommend to go through the full range of **Unsupervised Learning** algorithms and compare the performance of each resulting network structure to select the best structure.

The objective is to increase our chances of finding the optimal network for our purposes. Given that the number of possible networks grows super-exponentially with the number of nodes (Figure 8.11), this is a major challenge.

Number of Nodes	Number of Possible Networks
1	1
2	3
3	25
4	543
5	29281
6	3.7815×10^6
7	1.13878×10^9
8	7.83702×10^{11}
9	1.21344×10^{15}
10	4.1751×10^{18}
...	...
47	8.98454×10^{376}

Figure 8.11

It may not be immediately apparent how such an astronomical number of networks could be possible. Figure 8.12 displays how as few as 3 nodes can be combined in 25 different ways to form a network.

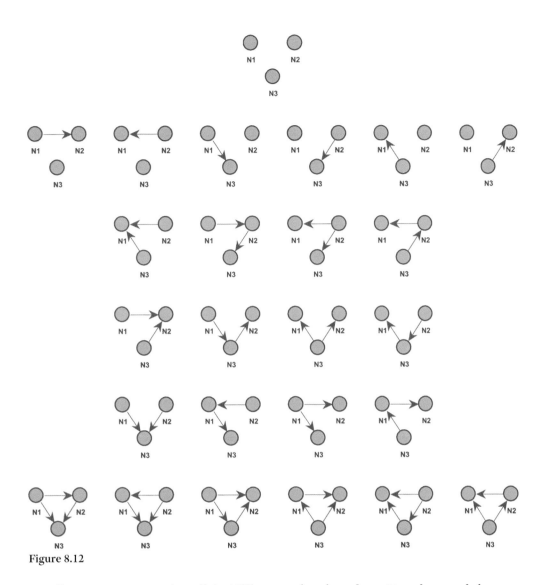
Figure 8.12

Needless to say, generating all 9×10^{376} networks—based on 47 nodes—and then selecting the best one is completely intractable.[4] An exhaustive search would only be feasible for a few nodes.

As a result, we have to use heuristic search algorithms to explore a small part of this huge space in order to find a *local* optimum. However, a heuristic search algorithm does not guarantee to find the *global* optimum. This is why BayesiaLab offers a range of distinct learning algorithms, which all use different search spaces or search strategies or both:

---

[4] For reference, it is estimated that there are between 10^{78} to 10^{82} atoms in the known, observable universe.

- Bayesian networks for **MWST** and **Taboo**.
- Essential Graphs for **EQ** and **SopLEQ**, i.e. graphs with edges and arcs representing classes of equivalent networks.
- Order of nodes for **Taboo Order**.

This diversity increases the probability of finding a solution close to the global optimum. Given adequate time and resources for learning, we recommend to employ the algorithms in the following sequence to find the best solution:

- **Maximum Weight Spanning Tree + Taboo**
- **Taboo** ("from scratch")
- **EQ** ("from scratch") + **Taboo**
- **SopLEQ + Taboo**
- **Taboo Order + Taboo**

However, to keep the presentation compact, we only illustrate the learning steps for the **EQ** algorithm (Figure 8.13): **Learning > Unsupervised Structural Learning > EQ**.

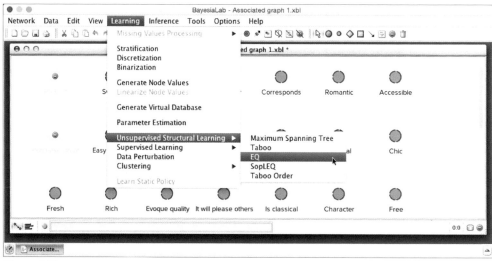

Figure 8.13

The network generated by the **EQ** algorithm is shown in Figure 8.14.

Chapter 8

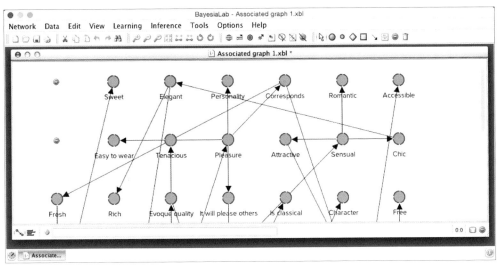

Figure 8.14

Pressing P, and then clicking the **Best-Fit** icon (), provides an interpretable view of the network (Figure 8.15). Additionally, rotating the network graph with the **Rotate Left** () and **Rotate Right** buttons () can help setting a suitable view.

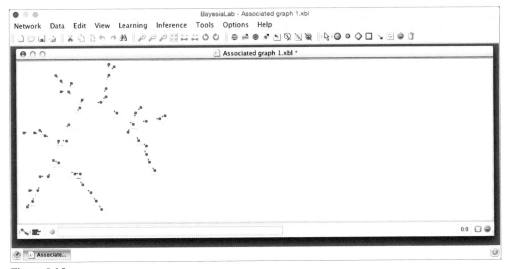

Figure 8.15

We now need to assess the quality of this network. Each of BayesiaLab's **Unsupervised Learning** algorithms uses the **MDL** score—internally—as the measure to optimize while searching for the best possible network. However, we can also employ the **MDL** score for explicitly rating the quality of a network.

Minimum Description Length

The **Minimum Description Length (MDL)** score is a two-component score, which has to be minimized to obtain the best solution. It has been used traditionally in the artificial intelligence community for estimating the number of bits required for representing (1) a "model," and (2) "data given this model."

In our machine-learning application, the "model" is a Bayesian network, consisting of a graph and probability tables. The second term is the log-likelihood of the data given the model, which is inversely proportional to the probability of the observations (data) given the Bayesian network (model). More formally, we write this as:

$$MDL(B,D) = \alpha DL(B) + DL(D \mid B), \tag{8.1}$$

where:

- α represents BayesiaLab's **Structural Coefficient** (the default value is 1), a parameter that permits changing the weight of the structural part of the MDL Score (the lower the value of α, the greater the complexity of the resulting networks),
- $DL(B)$ the number of bits to represent the Bayesian network B (graph and probabilities), and
- $DL(D|B)$ the number of bits to represent the dataset D given the Bayesian network B (likelihood of the data given the Bayesian network).

The minimum value for the first term, $DL(B)$, is obtained with the simplest structure, i.e. the fully unconnected network, in which all variables are stated as independent. The minimum value for the second term, $DL(D|B)$, is obtained with the fully connected network, i.e. a network corresponding to the analytical form of the joint probability distribution, in which no structural independencies are stated.

Thus, minimizing this score consists in finding the best trade-off between both terms. For a learning algorithm that starts with an unconnected network, the objective is to add a link for representing a probabilistic relationship if, and only if, this relationship reduces the log-likelihood of the data, i.e. $DL(D|B)$, by a large enough amount to compensate for the increase in the size of the network representation, i.e. $DL(B)$.

MDL Score Comparison

We now use the **MDL** score to compare the results of all learning algorithms.[5] We can look up the **MDL** score of the current network by pressing [W] while hovering with the cursor over the **Graph Panel**. This brings up an info box that reports a number of measures, including the **MDL** score, which is displayed here as "Final Score" (Figure 8.16).

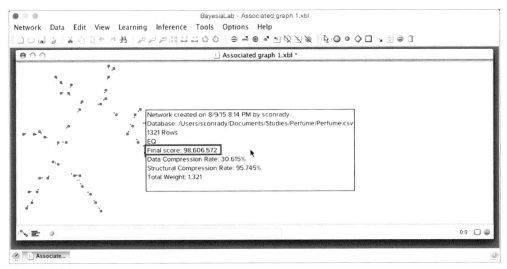

Figure 8.16

Alternatively, we can open up the **Console** via **Options** > **Console** > **Open Console** (Figure 8.17)

---

5   The **MDL** score can only be compared for networks with precisely the same representation of all variables, i.e. with the same discretization thresholds and the same data.

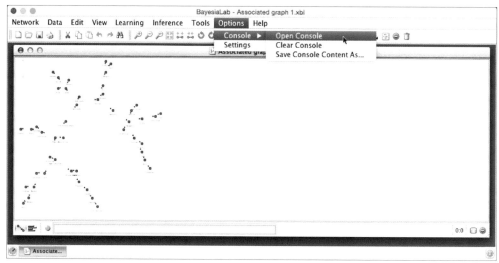

Figure 8.17

The **Console** maintains a kind of "log" that keeps track of the learning progress by recording the **MDL** score (or "Network Score") at each step of the learning process (Figure 8.18). Here, "Final Score" marks the **MDL** score of the current network, which is what we need to select the network with the lowest value.

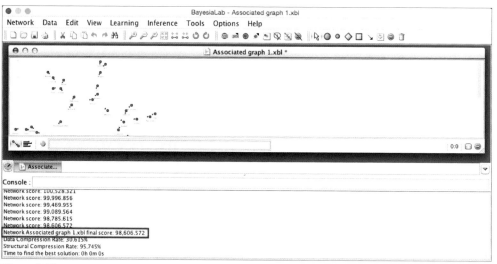

Figure 8.18

The **EQ** algorithm produces a network with an **MDL** score of 98,606.572. As is turns out, this performance is on par with all the other algorithms we considered, although we skip presenting the details in this chapter. Given this result, we can proceed with the **EQ**-learned network to the next step.

Data Perturbation

As a further safeguard against utilizing a sub-optimal network, BayesiaLab offers **Data Perturbation**, which is an algorithm that adds random noise (from within the interval [-1,1]) to the weight of each observation in the database.

In the context of our learning task, **Data Perturbation** can help escape from local minima, which we could have encountered during learning. We start this algorithm by selecting **Learning > Data Perturbation** (Figure 8.19)

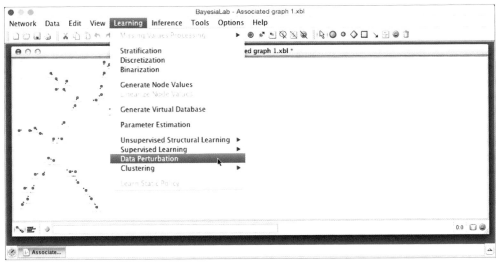

Figure 8.19

For **Data Perturbation**, we need to set a number of parameters (Figure 8.20).
The additive noise is always generated from a Normal distribution with a mean of 0, but we need to set the **Initial Standard Deviation**. A **Decay Factor** defines the exponential attenuation of the standard deviation with each iteration.

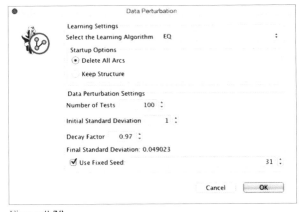

Figure 8.20

Upon completion of **Data Perturbation**, we see the newly learned network in the **Graph Panel**. Once again, we can retrieve the score by pressing [W] while hovering with the cursor over the **Graph Panel** or by looking it up in the **Console**. The score remains unchanged at 98,606.572. We can now be reasonably confident that we have found the optimal network given the original choice of discretization, i.e. the most compact representation of the joint probability distribution defined by the 47 manifest variables.

On this basis, we now switch to the **Validation Mode** (≡ or [F5]). Instead of examining individual nodes, however, we proceed directly to **Variable Clustering**.

Step 2: Variable Clustering

BayesiaLab's **Variable Clustering** is a hierarchical agglomerative clustering algorithm that uses **Arc Force** (i.e. the **Kullback-Leibler Divergence**) for computing the distance between nodes.

▶ Network Analysis in Chapter 7, p. 179.

At the start of **Variable Clustering**, each manifest variable is treated as a distinct cluster. The clustering algorithm proceeds iteratively by merging the "closest" clusters into a new cluster. Two criteria are used for determining the number of clusters:

▶ Arc Force in Chapter 7, p. 181.

- **Stop Threshold**: a minimum **Arc Force** value, below which clusters are not merged (a kind of significance threshold).
- **Maximum Cluster Size**: the maximum number of variables per cluster.

These criteria can be set via **Options > Settings > Learning > Variable Clustering** (Figure 8.21):

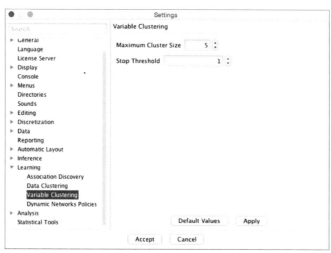

Figure 8.21

Whereas we do not advise to change the **Stop Threshold**, **Maximum Cluster Size** is more subjective. For building PSEMs, we recommend a value between 5 and 7, for reasons that will become clear when we show how latent variables are generated. If, however, the goal of **Variable Clustering** is dimensionality reduction, we suggest to increase **Maximum Cluster Size** to a much higher value, thus effectively eliminating it as a constraint.

The **Variable Clustering** algorithm can be started via **Learning > Clustering > Variable Clustering** or by using the shortcut ⬚S⬚.

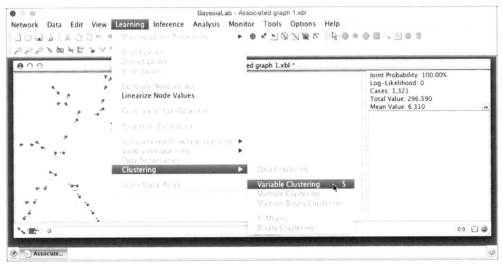

Figure 8.22

In this example, BayesiaLab identified 15 clusters, and each node is now color-coded according to its cluster membership. Figure 8.23 shows the standalone graph, outside the BayesiaLab window, for better legibility.

Figure 8.23

BayesiaLab offers several tools for examining and editing the proposed cluster structure. They are accessible from an extended menu bar (highlighted in Figure 8.24).

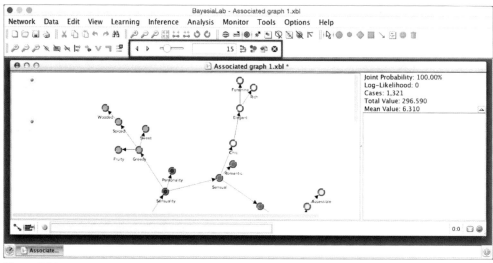

Figure 8.24

Dendrogram

The **Dendrogram** allows us to review the linkage of nodes within variable clusters. It can be activated via the corresponding icon ( ) in the extended menu. The lengths of the branches in the **Dendrogram** are proportional to the **Arc Force** between clusters.

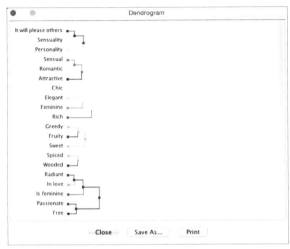

Figure 8.25

Also, the **Dendrogram** can be copied directly as a PDF or as a bitmap by right-clicking on it. Alternatively, it can be exported in various format via the **Save As...** button. As such, it can be imported into documents and presentation (Figure 8.26). This ability to copy and paste graphics applies to most graphs, plots, and charts in BayesiaLab.

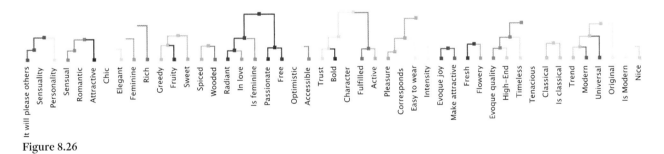

Figure 8.26

Cluster Mapping

As an alternative to **Dendrogram**, **Mapping** offers an intuitive approach to examining the just discovered cluster structure (Figure 8.27). It can be activated via the **Mapping** button in the menu bar ( ).

▸ Mapping in Chapter 6, p. 160.

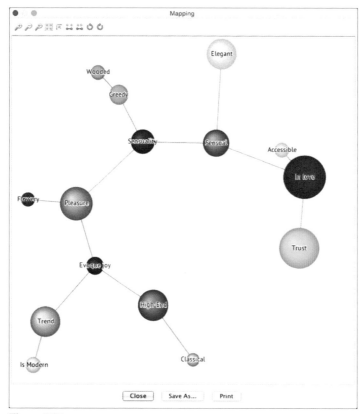

Figure 8.27

By hovering over any of the cluster "bubbles" with the cursor, BayesiaLab displays a list of all manifest nodes that are connected to that particular cluster (Figure 8.28). Each list of manifest variables is sorted according to the intra-cluster **Node Force**. This also explains the names displayed on the clusters. By default, each cluster takes on the name of the strongest manifest variables.

Figure 8.28

Number of Clusters

As explained earlier, BayesiaLab uses two criteria to determine the default number of clusters. We can change this number via the selector in the menu bar (Figure 8.29).

▸ Step 2: Variable Clustering, p. 216.

Figure 8.29

Both the **Dendrogram** and the **Mapping** view respond dynamically to any changes to the numbers of clusters.

Cluster Validation

The result of the **Variable Clustering** algorithm is purely descriptive. Once the question regarding the number of clusters is settled, we need to formally confirm our choice by clicking the **Validate Clustering** button (🗇) in the toolbar. Only then we trigger the creation of one **Class** per **Cluster**. At that time, all nodes become associated with unique **Classes**, which are named "*[Factor_i]*", with *i* representing the identifier of the factor. Additionally, we are prompted to confirm that we wish to keep the node colors that were generated during clustering (Figure 8.30).

Figure 8.30

The clusters are now saved, and the color coding is formally associated with the nodes. A **Clustering Report** (Figure 8.31) provides a formal summary of the new factors and their associated manifest variables.[6]

---

6 Note that we use the following terms interchangeably: "derived concept", "unobserved latent variable", "hidden cause", and "extracted factor".

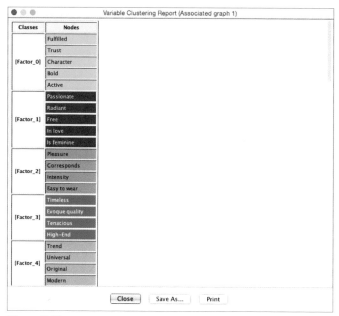

Figure 8.31

The **Class** icon ( ) in the lower right-hand corner of the window confirms that classes have been created corresponding to the factors. This concludes Step 2 and we formally close **Variable Clustering** via the stop icon ( ) on the extended toolbar.

Editing Factors

Beyond our choice with regard to the number of clusters, we also have the option of using our domain knowledge to modify which manifest nodes belong to specific factors. This can be done by right-clicking on the **Graph Panel**, and selecting **Edit Classes** and then **Modify** from the **Contextual Menu** (Figure 8.32). Alternatively, we can click the **Class** icon ( ). In our example, however, we show the **Class Editor** just for reference (Figure 8.33) as we keep all the original variable assignments in place.

Chapter 8

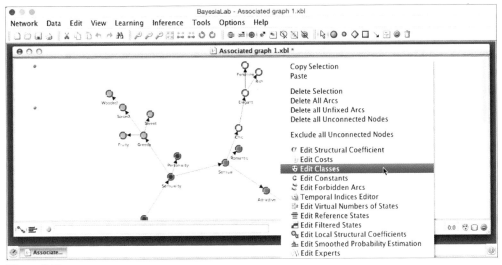

Figure 8.32

Figure 8.33

Cluster Cross-Validation

We now examine the robustness of the identified factors, i.e. how these factors respond to changes in sampling. This is particularly important for studies that are regularly repeated with new data, e.g. annual customer satisfaction surveys. Inevitably, survey samples are going to be different between the years. As a result, machine learning will probably discover somewhat different structures each time and, therefore, identify different clusters of nodes. Therefore, it is important to distinguish between a sampling artifact and a substantive change in the joint probability distribution. The latter, in the context of our example, would reveal a structural change in consumer behavior.

We start the validation process via **Tools > Cross-Validation > Variable Clustering > Data Perturbation** (Figure 8.34).

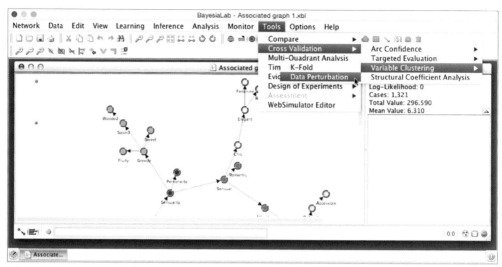

Figure 8.34

This brings up the dialogue box shown in Figure 8.35.

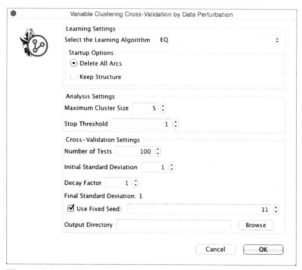

Figure 8.35

These settings specify that BayesiaLab will learn 100 networks with **EQ** and perform **Variable Clustering** on each one of them, all while maintaining the constraint of a maximum of 5 nodes per cluster and without any attenuation of the perturbation. Upon completion, we obtain a report panel (Figure 8.36), from which we initially select **Variable Clustering Report**.

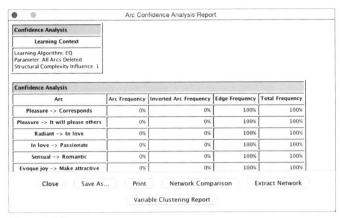

Figure 8.36

The **Variable Clustering Report** consists primarily of two large tables. The first table (Figure 8.37) in the report shows the cluster membership of each node in each network (only the first 12 columns are shown). Here, thanks to the colors, we can easily detect whether nodes remain clustered together between iterations or whether they "break up."

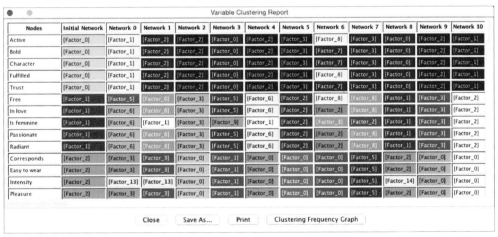

Figure 8.37

The second table (Figure 8.38) shows how frequently individual nodes are clustered together.

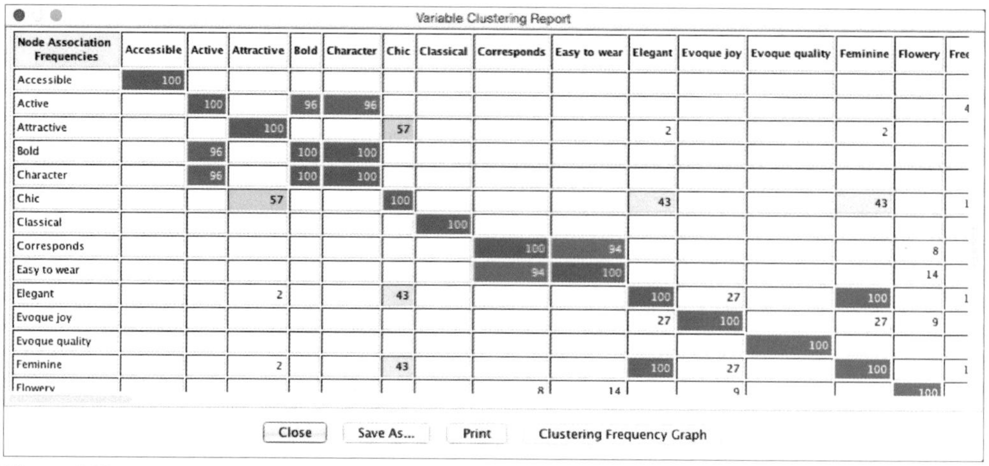

Node Association Frequencies	Accessible	Active	Attractive	Bold	Character	Chic	Classical	Corresponds	Easy to wear	Elegant	Evoque joy	Evoque quality	Feminine	Flowery	Frec
Accessible	100														
Active		100		96	96										4
Attractive			100			57				2			2		
Bold		96		100	100										
Character		96		100	100										
Chic			57			100				43			43		1
Classical							100								
Corresponds								100	94					8	
Easy to wear								94	100					14	
Elegant			2			43				100	27		100		1
Evoque joy										27	100		27	9	
Evoque quality												100			
Feminine			2			43				100	27		100		1
Flowery								8	14		9			100	

Figure 8.38

The **Clustering Frequency Graph** (Figure 8.39) provides yet another visualization of the clustering patterns. The thickness of the lines is proportional to the frequency of nodes being in the same cluster. Equally important for interpretation is the absence of lines between nodes. For instance, the absence of a line between *Flowery* and *Modern* says that they have never been clustered together in any of the 100 samples. If they were to cluster together in future iteration with new survey data, it would probably reflect a structural change in the market rather than a data sampling artifact.

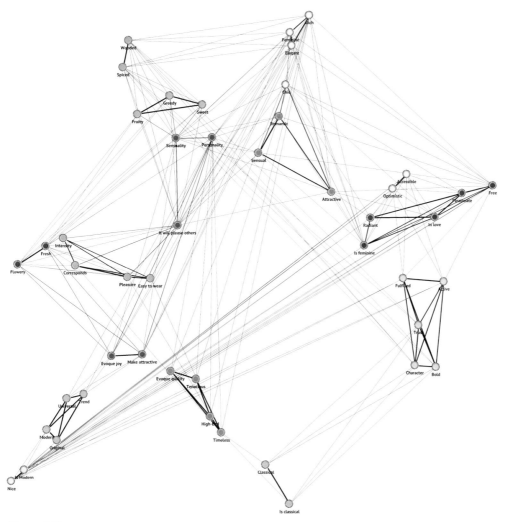

Figure 8.39

Step 3: Multiple Clustering

The **Cluster Cross-Validation** was merely a review, and it has not changed the factors that we confirmed when we clicked the **Validate Clustering** button (). Although we have defined these factors now in terms of classes of manifest variables, we still need to create the corresponding latent variables via **Multiple Clustering**. This process creates one discrete factor for each cluster of variables by performing data clustering on each subset of clustered manifest variables.

In traditional statistics, deriving such latent variables or factors is typically performed by means of Factor Analysis, e.g. Principal Components Analysis (PCA).

Data Clustering

Before we run this automatically across all factors with the **Multiple Clustering** algorithm, we will demonstrate the process on a single cluster of nodes, namely the nodes associated with *Factor_0*: *Active*, *Bold*, *Character*, *Fulfilled*, and *Trust*. We simply delete all other nodes and arcs and save this subset of nodes as a new, separate **xbl** file.

Figure 8.40

The objective of BayesiaLab's **Data Clustering** algorithm is to create a node that compactly represent the joint probability distribution defined by the variables of interest. We start **Data Clustering** via **Learning > Clustering > Data Clustering** (Figure 8.41). Unless we select a subset, **Data Clustering** will be applied to all nodes.

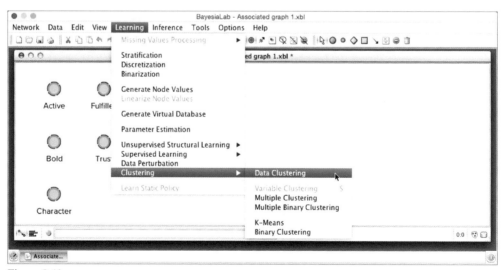

Figure 8.41

In the **Data Clustering** dialogue box we set the options as shown in Figure 8.42. Among the settings, we need to point out that we leave the number of states of the to-be-induced factor open; we only set a range, i.e. 2–5. This means that we let BayesiaLab determine the optimal number of states for representing the joint probability distribution.

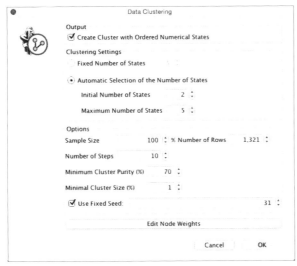

Figure 8.42

Upon completion of the clustering process, we obtain a graph with newly induced *[Factor_0]* being connected to all its associated manifest variables.

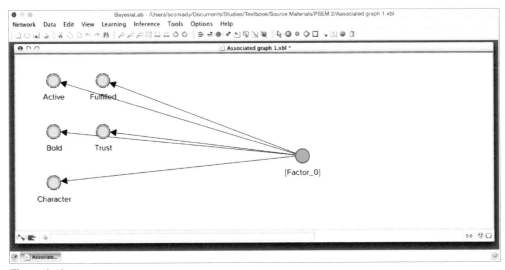

Figure 8.43

Furthermore, BayesiaLab produces a window that contains a comprehensive clustering report (Figure 8.44). Given its size, we split the full report across two pages (Figure 8.45 and Figure 8.46).

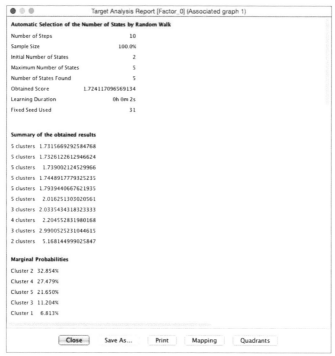

Figure 8.44

Chapter 8

Automatic Selection of the Number of States by Random Walk	
Number of Steps	10
Sample Size	100.00%
Initial Number of States	2
Maximum Number of States	5
Number of States Found	5
Obtained Score	1.724117097
Learning Duration	0h 0m 2s
Fixed Seed Used	31

Summary of the obtained results

5 clusters	1.731566929
5 clusters	1.732612261
5 clusters	1.739002125
5 clusters	1.744891778
5 clusters	1.793944067
5 clusters	2.016251303
3 clusters	2.033543432
4 clusters	2.204552832
3 clusters	2.990052523
2 clusters	5.168144999

Marginal Probabilities

Cluster 2	32.85%
Cluster 4	27.48%
Cluster 5	21.65%
Cluster 3	11.20%
Cluster 1	6.81%

Clustering Average Purity: 93.656%

Cluster	Purity	Neighborhood
Cluster 1	96.84%	Cluster 3 2.997% Cluster 5 0.111%
Cluster 5	96.51%	Cluster 2 3.444%
Cluster 2	92.97%	Cluster 4 4.699% Cluster 5 2.331%
Cluster 3	92.14%	Cluster 4 5.894% Cluster 1 1.965%
Cluster 4	92.06%	Cluster 2 5.360% Cluster 3 2.436%

Performance Indices
Contingency Table Fit 84.882%
Deviance 1,046.581
Hypercube Cells Per State 530.512

Figure 8.45

Node significance with respect to the information gain brought by the node to the knowledge [Factor_0]

Node	Mutual information	Normalized Mutual Information (%)	Relative significance	Mean Value	G-test	Degrees of Freedom	p-value	G-test (Data)	Degrees of Freedom (Data)	p-value (Data)
Trust	1.257	58.84%	1	6.791	2,301.14	16	0.00%	2,301.14	16	0.00%
Bold	1.098	51.44%	0.874	6.477	2,011.62	16	0.00%	2,011.62	16	0.00%
Fulfilled	1.051	49.22%	0.836	6.75	1,924.65	16	0.00%	1,924.65	16	0.00%
Active	0.995	46.59%	0.792	6.768	1,822.05	16	0.00%	1,822.05	16	0.00%
Character	0.837	39.18%	0.666	6.492	1,532.04	16	0.00%	1,532.04	16	0.00%

Node significance with respect to the information gain brought by the node to the knowledge of the target state

[Factor_0]= C2 (7,552) (32.854%)

Node	Binary mutual information	Normalized Binary Mutual Information (%)	Binary relative significance	Mean Value	Modal Value		A Priori Modal Value		G-test	Bayes Factor	Maximal Positive Variation		Maximal Negative Variation	
Trust	0.432	47.31%	1	7.661	<=8.2 (4/5)	88.25%	<=8.2 (4/5)	37.85%		1.221	<=8.2 (4/5)	50.40%	<=6.4 (3/5)	23.06%
Bold	0.333	36.44%	0.77	7.358	<=8.2 (4/5)	79.95%	<=8.2 (4/5)	36.03%		1.15	<=8.2 (4/5)	43.92%	<=6.4 (3/5)	13.13%
Fulfilled	0.307	33.58%	0.71	7.652	<=8.2 (4/5)	77.65%	<=8.2 (4/5)	35.96%		1.111	<=8.2 (4/5)	41.69%	<=6.4 (3/5)	17.38%
Active	0.29	31.76%	0.671	7.716	<=8.2 (4/5)	74.65%	<=8.2 (4/5)	34.90%		1.097	<=8.2 (4/5)	39.76%	<=6.4 (3/5)	18.30%
Character	0.24	26.32%	0.556	7.297	<=8.2 (4/5)	68.20%	<=8.2 (4/5)	32.78%		1.057	<=8.2 (4/5)	35.43%	>8.2 (5/5)	10.81%

[Factor_0]= C4 (5,934) (27.479%)

Node	Binary mutual information	Normalized Binary Mutual Information (%)	Binary relative significance	Mean Value	Modal Value		A Priori Modal Value			Bayes Factor	Maximal Positive Variation		Maximal Negative Variation	
Trust	0.338	39.86%	1	5.897	<=6.4 (3/5)	71.07%	<=8.2 (4/5)	37.85%		1.487	<=6.4 (3/5)	45.72%	>8.2 (5/5)	22.66%
Bold	0.298	35.15%	0.882	5.727	<=6.4 (3/5)	70.25%	<=8.2 (4/5)	36.03%		1.419	<=6.4 (3/5)	43.98%	>8.2 (5/5)	18.60%
Active	0.284	33.48%	0.84	6.109	<=6.4 (3/5)	65.84%	<=8.2 (4/5)	35.96%		1.398	<=6.4 (3/5)	40.86%	>8.2 (5/5)	22.55%
Fulfilled	0.275	32.37%	0.812	6.095	<=6.4 (3/5)	64.74%	<=8.2 (4/5)	35.96%		1.387	<=6.4 (3/5)	39.98%	>8.2 (5/5)	22.12%
Character	0.16	18.87%	0.473	5.806	<=6.4 (3/5)	55.92%	<=8.2 (4/5)	32.78%		1.137	<=6.4 (3/5)	30.49%	>8.2 (5/5)	17.37%

[Factor_0]= C5 (9,128) (21.650%)

Node	Binary mutual information	Normalized Binary Mutual Information (%)	Binary relative significance	Mean Value	Modal Value		A Priori Modal Value			Bayes Factor	Maximal Positive Variation		Maximal Negative Variation	
Trust	0.517	68.65%	1	9.409	>8.2 (5/5)	91.61%	<=8.2 (4/5)	37.85%		1.998	>8.2 (5/5)	68.67%	<=8.2 (4/5)	29.46%
Bold	0.405	53.68%	0.782	8.942	<=4.6 (2/5)	76.92%	<=8.2 (4/5)	36.03%		2.006	>8.2 (5/5)	57.77%	<=6.4 (3/5)	23.82%
Fulfilled	0.385	51.14%	0.745	9.146	>8.2 (5/5)	84.97%	<=8.2 (4/5)	35.96%		1.838	>8.2 (5/5)	61.20%	<=8.2 (4/5)	25.82%
Character	0.351	46.57%	0.678	8.992	>8.2 (5/5)	78.67%	<=8.2 (4/5)	34.90%		1.817	>8.2 (5/5)	56.34%	<=6.4 (3/5)	21.59%
Active	0.346	45.95%	0.669	9.042	>8.2 (5/5)	83.92%	<=8.2 (4/5)	32.78%		1.761	>8.2 (5/5)	59.16%	<=8.2 (4/5)	25.46%

[Factor_0]= C3 (3,941) (11.204%)

Node	Binary mutual information	Normalized Binary Mutual Information (%)	Binary relative significance	Mean Value	Modal Value		A Priori Modal Value			Bayes Factor	Maximal Positive Variation		Maximal Negative Variation	
Bold	0.252	49.78%	1	3.752	<=4.6 (2/5)	68.92%	<=8.2 (4/5)	36.03%		2.732	<=4.6 (2/5)	58.55%	<=8.2 (4/5)	34.68%
Fulfilled	0.241	47.60%	0.956	3.955	<=4.6 (2/5)	58.11%	<=8.2 (4/5)	35.96%		2.764	<=4.6 (2/5)	49.55%	<=8.2 (4/5)	35.28%
Trust	0.222	43.88%	0.881	4.309	<=4.6 (2/5)	49.32%	<=8.2 (4/5)	37.85%		2.733	<=4.6 (2/5)	61.20%	<=8.2 (4/5)	25.82%
Active	0.216	42.76%	0.859	4.029	<=4.6 (2/5)	56.08%	<=8.2 (4/5)	34.90%		2.675	<=4.6 (2/5)	47.30%	<=8.2 (4/5)	32.20%
Character	0.183	36.19%	0.727	3.632	<=4.6 (2/5)	52.03%	<=8.2 (4/5)	32.78%		2.306	<=4.6 (2/5)	41.51%	<=8.2 (4/5)	30.75%

[Factor_0]= C1 (2,022) (6.813%)

Node	Binary mutual information	Normalized Binary Mutual Information (%)	Binary relative significance	Mean Value	Modal Value		A Priori Modal Value			Bayes Factor	Maximal Positive Variation		Maximal Negative Variation	
Trust	0.26	72.47%	1	1.962	<=2.8 (1/5)	83.33%	<=8.2 (4/5)	37.85%		3.695	<=2.8 (1/5)	76.90%	<=8.2 (4/5)	35.63%
Bold	0.254	70.79%	0.977	1.899	<=2.8 (1/5)	92.22%	<=8.2 (4/5)	36.03%		3.496	<=2.8 (1/5)	84.05%	<=8.2 (4/5)	36.03%
Fulfilled	0.239	66.58%	0.919	2.027	<=2.8 (1/5)	82.22%	<=8.2 (4/5)	35.96%		3.561	<=2.8 (1/5)	75.26%	<=8.2 (4/5)	34.85%
Active	0.227	63.23%	0.872	2.126	<=2.8 (1/5)	76.67%	<=8.2 (4/5)	34.90%		3.541	<=2.8 (1/5)	70.08%	<=8.2 (4/5)	32.68%
Character	0.189	52.79%	0.728	2.137	<=2.8 (1/5)	81.11%	<=8.2 (4/5)	32.78%		3.183	<=2.8 (1/5)	72.18%	<=8.2 (4/5)	31.67%

Figure 8.46

Relative Significance

In the second part of the report (Figure 8.46), the variables are sorted by **Relative Significance** with respect to the **Target Node**, which is *[Factor_0]*.

$$RS_i = \frac{I(M_i, F)}{\max_j I(M_j, F)}, \tag{8.2}$$

where M_i represents the ith manifest variable, and F represents the factor variable. The function $I(\cdot)$ computes the **Mutual Information**.

Mapping

From the window that contains the report (Figure 8.44), we can also produce a **Mapping** of the clusters (Figure 8.47).

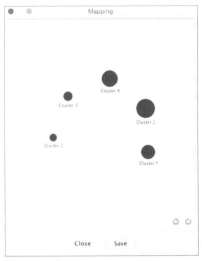

Figure 8.47

This graph displays three properties of the identified states (Cluster 1–Cluster 5) within the new factor node:

- The saturation of the blue represents the purity of the clusters: the higher the purity, the higher the saturation of the color. Here, all purities are in the 90%+ range, which is why they are all deep blue.
- The sizes represent the respective marginal probabilities of the states (clusters). We will see this distribution again once we open the **Monitor** of the new factor node.
- The distance between any two clusters is proportional to the neighborhood of the clusters.

Quadrants

Clicking the **Quadrants** button in the report window (Figure 8.44) brings up the options for graphically displaying the relative importance of the node with regard to the induced factor (Figure 8.48).

Figure 8.48

For our example, we select **Mutual Information**. Furthermore, we do not need to normalize the means as all values of the nodes in this **Cluster** are recorded on the same scale.

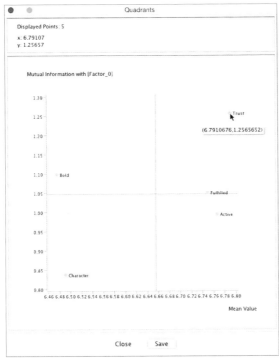

Figure 8.49

This **Quadrant Plot** highlights two measures that are relevant for interpretation:
- **Mutual Information** on the y-axis, i.e. the importance of each manifest variables with regard to the latent variable, *[Factor_0]*.
- The **Mean Value** of each manifest variable on the x-axis.

This plot shows us that the most important variable is I(*Trust,[Factor_0]*)=1.26. It is also the variable with the highest expected satisfaction level, i.e. E(*Trust*)=6.79.

When hovering with the cursor over the plot, the upper panel of the **Quadrant Plot** window returns the exact coordinates of the respective point, i.e. **Mutual Information** and **Mean Value** in this example.

Upon closing the **Quadrant Plot** and the report window, we return to the **Graph Panel**. It shows the newly induced **Factor** directly connected to all its associated manifest variables. Applying the **Automatic Layout** (P) produces a suitable view of the network produced by the **Data Clustering** process (Figure 8.50).

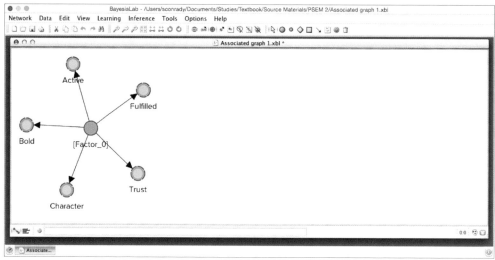

Figure 8.50

After switching to the **Validation Mode** (≡ or F5), we open the **Monitors** for all nodes. We can see five states (clusters) for *[Factor_0]*, labeled *C1* through *C5*, as well as their marginal distribution. This distribution was previously represented as the "bubble size" of Clusters 1–5 in Figure 8.47.

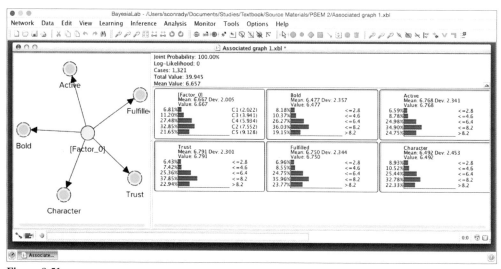

Figure 8.51

In the **Monitor** of *[Factor_0]*, we see that the name of each state carries a value shown in parentheses, e.g. *C1 (2.022)*. This value is the weighted average of the associated manifest variables given the state *C1*, where the weight of each variable is its **Relative Significance** with respect to *[Factor_0]*. That means that given the state *C1* of *[Factor_0]*, the weighted mean value of *Trust*, *Bold*, *Fulfilled*, *Active*, and *Character*

▶ Relative Significance, p. 233.

is 2.022. This becomes more apparent when we actually set the evidence *C1* (Figure 8.52).

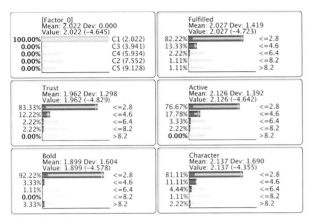

Figure 8.52

Given that all the associated manifest variables share the same satisfaction level scale, the values of the created states can also be interpreted as satisfaction levels. State *C1* summarizes the "low" ratings across the manifest nodes. Conversely, *C5* represents mostly the "high" ratings; the other states cover everything else in between.

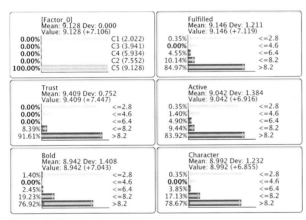

Figure 8.53

It is important to understand that each underlying record was assigned a specific state of *[Factor_0]*. In other words, the hidden variable is no longer hidden. It has been added to the database and imputed for all respondents. The imputation is done via **Maximum Likelihood**: given the satisfaction levels observed for each of the 5 manifest variables, the state with the highest posterior probability is assigned to the respondent.

We can easily verify this by scrolling through each record in the database. To do so, we first set *[Factor_0]* as target node by right-clicking on it and selecting **Set as**

Target Node from the contextual menu (Figure 8.54). Note that the **Monitor** corresponding to the **Target Node** turns red.

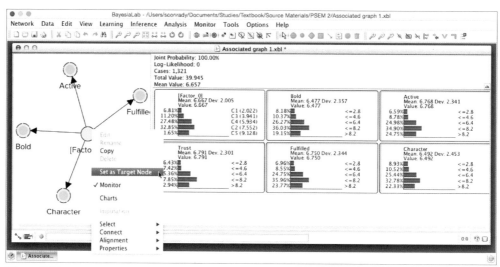

Figure 8.54

Then, we select **Inference** > **Interactive Inference** (Figure 8.55).

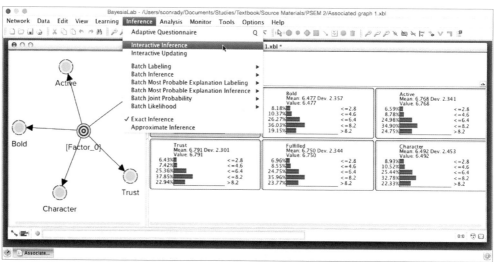

Figure 8.55

Using the record selector in the extended toolbar, we can now scroll through each record in the associated database. The **Monitors** of the manifest nodes show the actual survey observations, while the **Monitor** of *[Factor_0]* shows the posterior probability distribution of the states given these observations. The state highlighted in light blue the marks modal value, i.e. the "winning" state, which is the imputed state now

Chapter 8

recorded in the database (Figure 8.56). Clicking the **Stop Inference** icon (✖) closes this function.

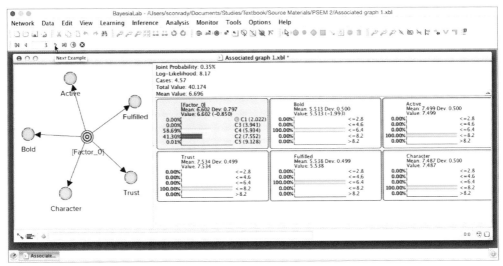

Figure 8.56

Network Performance Analysis

While the **Performance Indices** shown in the **Data Clustering Report** (Figure 8.45) have already included some measures of fit, we can further study this point by starting a more formal performance analysis: **Analysis > Network Performance > Overall** (Figure 8.57).

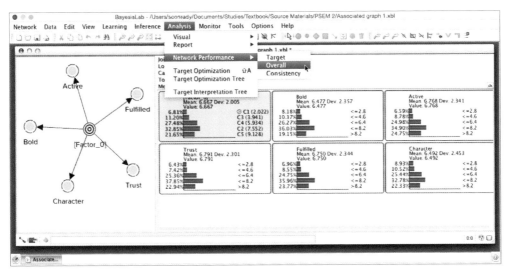

Figure 8.57

The resulting report provides us with measures of how well this network represents the underlying database (Figure 8.58).

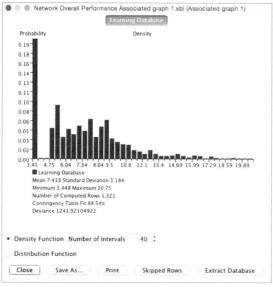

Figure 8.58

Contingency Table Fit

Of particular interest is BayesiaLab's **Contingency Table Fit (CTF)**, which measures the quality of the JPD representation. It is defined as:

$$CTF(B) = \frac{\overline{ll}(B_u) - \overline{ll}(B)}{\overline{ll}(B_f) - \overline{ll}(B)},$$ (8.3)

where:

$\overline{ll}(B)$ is the mean of the log-likelihood of the data given the network currently under study,

$\overline{ll}(B_u)$ is the mean of the log-likelihood of the data given the fully unconnected network, i.e. the "worst-case scenario," and

$\overline{ll}(B_f)$ is the mean of the log-likelihood of the data given the fully connected network, i.e. the "best-case scenario." The fully connected network is the complete graph, in which all nodes have a direct link to all other nodes. Therefore, it is the exact representation of the chain rule, without any conditional independence assumptions in the representation of the joint probability distribution.

Accordingly, we can interpret the following key values of the **CTF**:

- **CTF** is equal to 100 if the network represents the joint probability distribution of the data without any approximation, i.e. it has the same log-likelihood as the fully connected network.
- **CTF** is equal to 0 if the network represents a joint probability distribution no different than the one produced by the fully unconnected network, in which all the variables are marginally independent.

The main benefit of employing **CTF** as a quality measure is that it has normalized values ranging between 0% and 100%.[7]

CTF in Practice

It must be emphasized that **CTF** measures only the quality of the network in terms of its data fit. As such, it represents the second term in the definition of the **MDL** score ($MDL(B,D) = \alpha DL(B) + \overline{DL(D\mid B)}$). Even though this says, the higher the **CTF**, the better the representation of the JPD, we are *not* aiming for **CTF**=100%. This would conflict with the objective of finding a compact representation of the JPD.

▸ Minimum Description Length, p. 212.

The Naive structure of the network used for **Data Clustering** implies that the entire JPD representation relies on the **Factor** node. Removing this node would produce a fully unconnected network with a **CTF**=0%. Therefore, BayesiaLab excludes—but does not remove—the **Factor** node when computing the **CTF**. This allows measuring the quality of the JPD representation with the induced clusters only.

It is not easy to recommend a threshold value below which the **Factor** should be "reworked," as the **CTF** depends directly on the size of the JPD and the number of states of the **Factor**. For instance, given a **Factor** with 4 states and 2 binary manifest variables, a **CTF** any lower than 100% would be a poor representation of the JPD, as the JPD only consists of 4 cells. On the other hand, given 10 manifest variables, with 5 states each, and a **Factor** also consisting of 5 states, a **CTF** of 50% would be a very compact representation of the JPD. This means that 5 states would manage to represent a JPD of 5^{10} cells with a quality of 50%.

Returning to the context of our PSEM workflow, we have the following 3 conditions:

---

7   This measure can become negative if the parameters of the model is not estimated from the currently associated database.

1. A maximum number of 5 variables per cluster of variables;
2. **Manifest** variables with 5 states;
3. **Factors** with a maximum of 5 states.

In this situation, we recommend using 70% as an alert threshold. However, this threshold level would have to be reduced if conditions #1 and #2 increased in their values or if condition #3 decreased.

Multiple Clustering

The previous section on **Data Clustering** dealt exclusively with the induction of *[Factor_0]*. In our perfume study, however, we have 15 clusters of manifest variables, for which 15 **Factors** need to be induced. This means that all steps applicable to **Data Clustering** need to be repeated 15 times. BayesiaLab simplifies this task by offering the **Multiple Clustering** algorithm, which automates all necessary steps for all factors.

We now return to the original network, last presented in Figure 8.23. On its basis, we can immediately start **Multiple Clustering: Learning > Clustering > Multiple Clustering**.

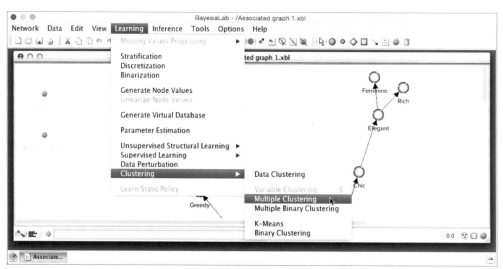

Figure 8.59

Compared to the dialogue box for **Data Clustering** (Figure 8.42), the options for **Multiple Clustering** are much expanded (Figure 8.60). Firstly, we need to specify an **Output Directory** for the to-be-learned networks. This will produce a separate network for each **Factor**, which we can subsequently examine. Furthermore, we want the new **Factors** to be connected to their manifest variables, but we do not wish the manifest variables to be connected amongst themselves. In fact, we have already learned the

Chapter 8

relationships between the manifest variables during Step 1. These relationships will ultimately be encoded via the connections between their associated **Factors** upon completion of Step 3. We consider these new **Factor** nodes belonging to the *second* layer of our hierarchical Bayesian network. This also means that, at this point, all structural learning involving the nodes of the *first* layer, i.e. the manifest variables, is completed.

We set the above requirements via **Connect Factors to their Manifest Variables** and **Forbid New Relations with Manifest Variables**. Another helpful setting is **Compute Manifests' Contributions to their Factor**, which helps to identify the dominant nodes within each factor.

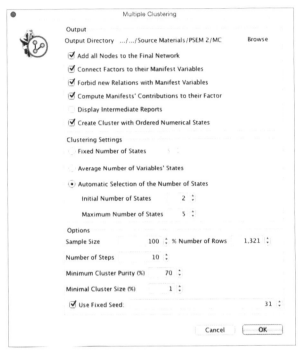

Figure 8.60

The **Multiple Clustering** process concludes with a report, which shows details regarding the generated clustering (Figure 8.61). Among the many available metrics, we can check the minimum value of the **Contingency Table Fit**, which is reported as 76.16%. Given the recommendations we provided earlier, this suggests that we did not lose too much information by inducing the latent variables.

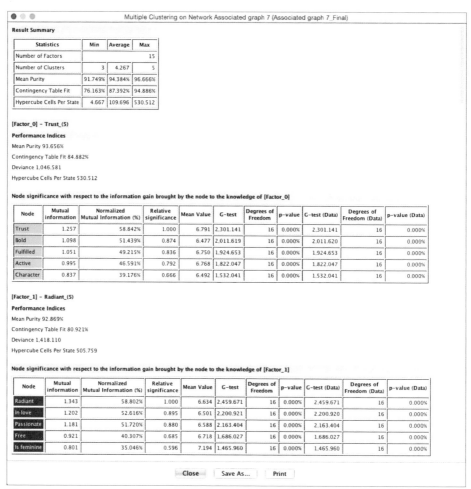

Figure 8.61

We can save the report or proceed straight to the new network in the **Graph Panel** (Figure 8.62), which has all nodes arranged in a grid-like arrangement: manifest variables are on the left; the new factors are stacked up on the right.

Chapter 8

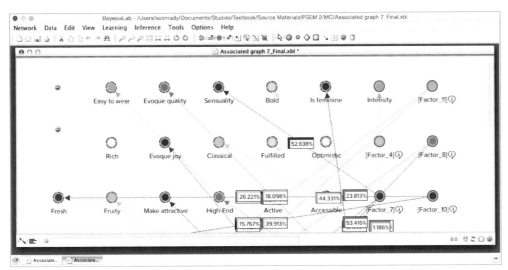

Figure 8.62

Upon applying **Automatic Layout** ([P]), we can identify 15 **Factors** surrounded by their manifest nodes, arranged almost like a field of flowers (Figure 8.63).

Figure 8.63

The **Arc Comments**, which are shown by default, display the **Contribution** of each manifest variable towards its **Factor**. Once we turn off the **Arc Comments** (🖼) and turn on the **Node Comments** (🖼), we see that the **Node Comments** contain the name of the "strongest" associated manifest variable, along with the number of associated manifest variables in parentheses. Figure 8.64 shows a subset of the nodes with their respective **Node Comments**.

Chapter 8

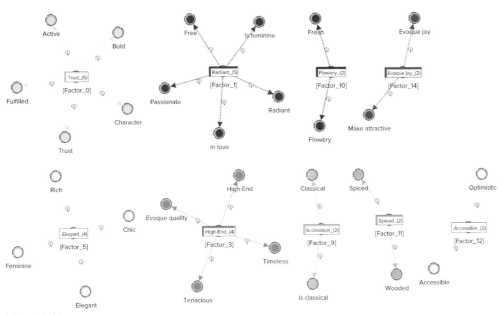

Figure 8.64

Also, by going into our previously specified output directory, we can see that 15 new sub-networks (in BayesiaLab's **xbl** format for networks) were generated (Figure 8.65). Any of these files would allow us to study the sub-networks' properties, as we did for the single factor that was generated by **Data Clustering**.

Additionally, one more file was created in this directory, which is highlighted in Figure 8.65. The file marked the suffix "_Final" is the network that consists of both the original manifest variables and the newly created **Factors**. As such, it is labeled as the "final" network in BayesiaLab parlance. It is also the network that is currently active.

Name	Date Modified	Size	Kind
Associated graph 1 [Factor_0].xbl	Aug 10, 2015, 6:12 PM	11 KB	BayesiaLab.app Document
Associated graph 1 [Factor_1].xbl	Aug 10, 2015, 6:12 PM	11 KB	BayesiaLab.app Document
Associated graph 1 [Factor_2].xbl	Aug 10, 2015, 6:12 PM	8 KB	BayesiaLab.app Document
Associated graph 1 [Factor_3].xbl	Aug 10, 2015, 6:12 PM	10 KB	BayesiaLab.app Document
Associated graph 1 [Factor_4].xbl	Aug 10, 2015, 6:12 PM	10 KB	BayesiaLab.app Document
Associated graph 1 [Factor_5].xbl	Aug 10, 2015, 6:12 PM	9 KB	BayesiaLab.app Document
Associated graph 1 [Factor_6].xbl	Aug 10, 2015, 6:12 PM	8 KB	BayesiaLab.app Document
Associated graph 1 [Factor_7].xbl	Aug 10, 2015, 6:12 PM	8 KB	BayesiaLab.app Document
Associated graph 1 [Factor_8].xbl	Aug 10, 2015, 6:12 PM	8 KB	BayesiaLab.app Document
Associated graph 1 [Factor_9].xbl	Aug 10, 2015, 6:12 PM	6 KB	BayesiaLab.app Document
Associated graph 1 [Factor_10].xbl	Aug 10, 2015, 6:12 PM	6 KB	BayesiaLab.app Document
Associated graph 1 [Factor_11].xbl	Aug 10, 2015, 6:12 PM	6 KB	BayesiaLab.app Document
Associated graph 1 [Factor_12].xbl	Aug 10, 2015, 6:12 PM	6 KB	BayesiaLab.app Document
Associated graph 1 [Factor_13].xbl	Aug 10, 2015, 6:12 PM	5 KB	BayesiaLab.app Document
Associated graph 1 [Factor_14].xbl	Aug 10, 2015, 6:13 PM	6 KB	BayesiaLab.app Document
Associated graph 1_Final.xbl	**Aug 10, 2015, 6:13 PM**	**78 KB**	**BayesiaLab.app Document**

Figure 8.65

In this context, BayesiaLab also created two new **Classes**:
- **Manifest**, which contains all the manifest variables;
- **Factor**, which contains all the latent variables.

Opening the **Class Editor** confirms their presence (highlighted items in Figure 8.66).

247

Figure 8.66

Step 4: Completing the Probabilistic Structural Equation Model

Based on the "final" network, we can proceed to the next step in our network building process. We now introduce *Purchase Intent*, which had been excluded up to this point. Clicking this node while holding [X] renders it "un-excluded." This makes *Purchase Intent* available for learning. Additionally, we designate *Purchase Intent* as **Target Node** by double-clicking the node while holding [T].

Looking for an SEM-type network structure stipulates that manifest variables be connected exclusively to the factors and that all the connections with *Purchase Intent* must go through the factors. We have already imposed this constraint by setting the option **Forbid New Relations with Manifest Variables** in the **Multiple Clustering** dialogue box (Figure 8.60). This created so-called **Forbidden Arcs**, which prevent learning algorithms from creating new arcs between the specified nodes. BayesiaLab indicates the presence of **Forbidden Arcs** with an icon in the lower right-hand corner of the **Graph Panel** window ( ). Clicking on the icon brings up the **Forbidden Arc Editor**, which allows us to review the currently set constraints (Figure 8.67). We see that the nodes belonging to the **Class Manifest** must not have any links to any other nodes, i.e. both directions are "forbidden."

Figure 8.67

Upon confirming these constraints, we start **Unsupervised Learning** to generate a network that includes the **Factors** and the **Target Node**. In this particular situation, we need to utilize **Taboo Learning**. It is the only algorithm in BayesiaLab that can learn a new structure on top of an existing network structure and simultaneously guarantee to keep **Fixed Arcs** unchanged.[8] This is important as the arcs linking the **Factors** and their manifest variables are such **Fixed Arcs**. To distinguish them visually, **Fixed Arcs** appear as dotted lines in the network, as opposed to the solid lines of "regular" arcs.

We start **Taboo Learning** from the main menu by selecting **Learning > Unsupervised Structural Learning > Taboo** (Figure 8.68) and check the option **Keep Network Structure** in the **Taboo Learning** dialogue box (Figure 8.69).

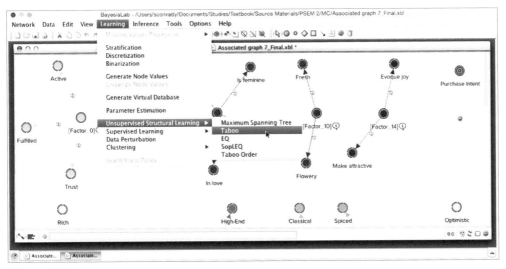

Figure 8.68

---

8   **EQ** can also be used for structural learning on top of an existing network, but as it searches in the space of Essential Graphs, there is no guarantee that the **Fixed Arcs** remain unchanged.

Figure 8.69

Upon completing the learning process, we obtain the network shown in Figure 8.70.

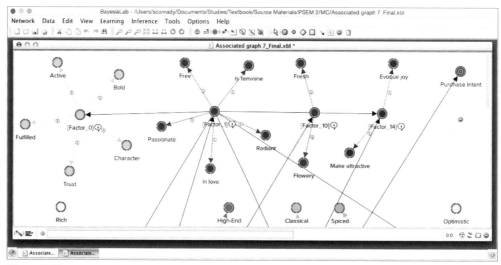

Figure 8.70

As in Step 1, we also try to improve the quality of this network by using the **Data Perturbation** algorithm.

Figure 8.71

As it turns out, this algorithm allowed us to escape from a local optimum and returned a final network with a lower **MDL** score. By using **Automatic Layout** (P) and turning on **Node Comments** (😀), we can quickly transform this network into a much more interpretable format (Figure 8.72).

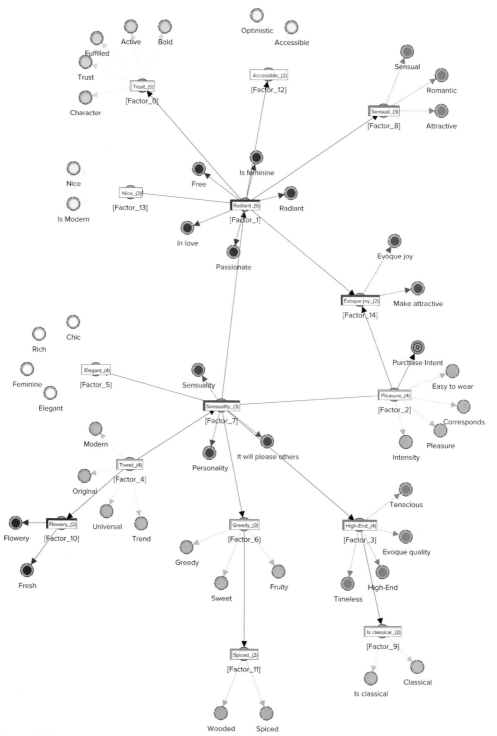

Figure 8.72

Now we see how the manifest variables are "laddering up" to the **Factors**, and we also see how the **Factors** are related to each other. Most importantly, we can observe

Chapter 8

where the *Purchase Intent* node was attached to the network during the learning process. The structure conveys that *Purchase Intent* is only connected to *[Factor_2]*, which is labeled with the **Node Comment** "*Pleasure_(4)*."

Key Drivers Analysis

Our **Probabilistic Structural Equation Model** is now complete, and we can use it to perform the analysis part of this exercise, namely to find out what "drives" *Purchase Intent*. We return to the **Validation Mode** (≡ or F5).

In order to understand the relationship between the factors and *Purchase Intent*, we want to "tune out" all manifest variables for the time being. We can do so by right-clicking the **Use-of-Classes** icon ( ) in the bottom right corner of the **Graph Panel** window. This brings up a list of all **Classes**. By default, all are checked and thus visible (Figure 8.73).

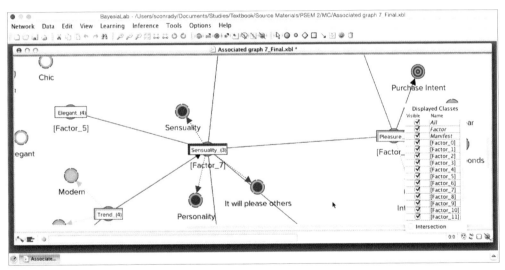

Figure 8.73

For our purposes, we want to un-check *All* and then only check the class *Factor* (Figure 8.74).

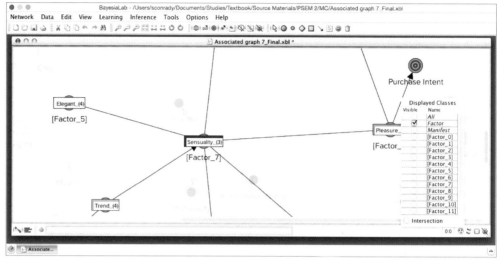

Figure 8.74

In the resulting view, all the **Manifest Nodes** are transparent, so the relationship between the **Factors** becomes visually more prominent. By de-selecting the manifest variables in this way, we also exclude them from the following visual analysis.

Target Analysis

In line with our objective of learning about the key drivers in this domain, we proceed to analyze the association of the newly created **Factors** with *Purchase Intent*.

We return to the **Validation Mode** (≡ or [F5]), in which we can use two approaches to learn about the relationships between **Factors** and the **Target Node**: we first perform a visual analysis and then generate a report in table format.

Visual Analysis

We initiate the visual analysis by selecting **Analysis > Visual > Target Mean Analysis > Standard** (Figure 8.75),

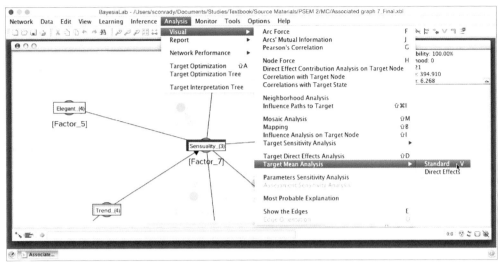

Figure 8.75

This brings up a dialogue box with options as shown in Figure 8.76. Given the context, selecting **Mean** for both the **Target Node** and the **Variables (Nodes)** is appropriate.

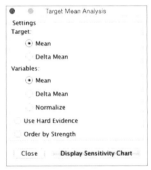

Figure 8.76

Upon clicking **Display Sensitivity Chart**, the resulting plot shows the response curves of the target node as a function of the values of the **Factors** (Figure 8.76). This allows an immediate interpretation of the strength of association.

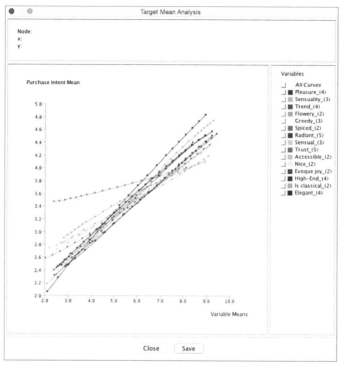

Figure 8.77

Target Analysis Report

As an alternative to the visual analysis, we now run the **Target Analysis Report: Analysis > Report > Target Analysis > Total Effects on Target**. Although "effects" carries a causal connotation, we need to emphasize that we are strictly examining associations. This means that we perform observational inference as we generate this report.

Chapter 8

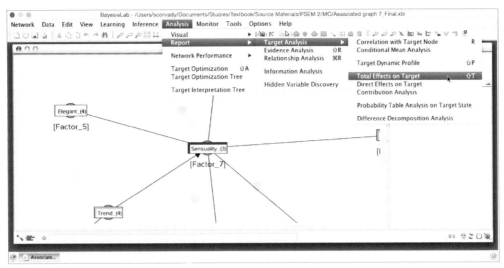

Figure 8.78

A new window opens up to present the report (Figure 8.79).[9]

Node	Comment	Value/Mean	Standardized Total Effects	Total Effects	G-test	Degrees of Freedom	p-value	G-test (Data)	Degrees of Freedom (Data)	p-value (Data)
[Factor_2]	Pleasure_(4)	5.884	0.786	0.399	1,262.892	15	0.000%	1,262.892	15	0.000%
[Factor_7]	Sensuality_(3)	6.140	0.670	0.368	789.448	20	0.000%	1,032.791	20	0.000%
[Factor_14]	Evoque joy_(2)	6.338	0.601	0.354	583.390	10	0.000%	867.273	10	0.000%
[Factor_5]	Elegant_(4)	6.648	0.549	0.333	477.220	15	0.000%	925.110	15	0.000%
[Factor_4]	Trend_(4)	6.267	0.547	0.335	477.545	20	0.000%	892.101	20	0.000%
[Factor_3]	High-End_(4)	6.139	0.547	0.327	474.937	20	0.000%	872.578	20	0.000%
[Factor_1]	Radiant_(5)	6.688	0.534	0.322	456.293	20	0.000%	736.732	20	0.000%
[Factor_8]	Sensual_(3)	6.524	0.437	0.256	290.453	15	0.000%	721.083	15	0.000%
[Factor_6]	Greedy_(3)	6.210	0.429	0.271	276.812	10	0.000%	505.669	10	0.000%
[Factor_0]	Trust_(5)	6.667	0.428	0.277	278.338	20	0.000%	677.105	20	0.000%
[Factor_13]	Nice_(2)	6.755	0.404	0.274	242.647	10	0.000%	595.617	10	0.000%
[Factor_12]	Accessible_(2)	6.626	0.386	0.256	225.823	20	0.000%	547.657	20	0.000%
[Factor_10]	Flowery_(2)	6.752	0.334	0.211	162.399	15	0.000%	502.436	15	0.000%
[Factor_9]	Is classical_(2)	6.248	0.273	0.186	112.598	20	0.000%	305.029	20	0.000%
[Factor_11]	Spiced_(2)	4.794	0.151	0.094	34.196	15	0.320%	164.995	15	0.000%

Figure 8.79

The **Total Effect (TE)** is estimated as the derivative of the **Target Node** with respect to the driver node under study.

$$\text{TE}(X, Y) = \frac{\delta_Y}{\delta_X}, \tag{8.4}$$

---

9   Under **Options > Settings > Reporting**, we can check **Display the Node Comments in Tables** so that **Node Comments** appears in addition to the **Node Names** in all reports.

where *X* is the analyzed variable and *Y* is the **Target Node**. The **Total Effect** represents the impact of a small modification of the mean of a driver node on the mean of the target node. The **Total Effect** is the ratio of these two values.

This way of measuring the effect of the **Factors** on the **Target Node** assumes the relationships to be locally linear. Even though this is not always a correct assumption, it can be reasonable for simulating small changes of satisfaction levels.

As per the report, *[Factor_2]* provides the strongest **Total Effect** with a value of 0.399. This means that observing an increase of one unit in the level of the concept represented by *[Factor_2]* leads to a posterior probability distribution for *Purchase Intent* that has expected value that is 0.399 higher compared to the marginal value.

The **Standardized Total Effect (STE)** is also displayed. It represents the **Total Effect** multiplied by the ratio of the standard deviation of the driver node and the standard deviation of the **Target Node**.

$$\text{STE}(X,Y) = \frac{\hat{\delta}_Y}{\hat{\delta}_X} \times \frac{\sigma_X}{\sigma_Y} \qquad (8.5)$$

This means that **STE** takes into account the "potential" of the driver under study.

In the report, the results are sorted by the **Standardized Total Effect** in descending order. This immediately highlights the order of importance of the **Factors** relative to the **Target Node**, *Purchase Intent*.

Independence Tests

In the columns further to the right in the report, the results of independence tests between the nodes are reported:

- Chi-Square (X^2) test or G-test: The independence test is computed on the basis of the network between each driver node and the target variable. It is possible to change the type of independence from Chi-Square (X^2) test to G-test via **Options > Settings > Statistical Tools**.
- Degree of Freedom: Indicates the degree of freedom between each driver node and the target node in the network.
- p-value: the p-value is the probability of observing a value as extreme as the test statistic by chance.

If a database is associated with the network, as is the case here, the independence test, the degrees of freedom, and the p-value are also computed directly from the underlying data.

Chapter 8

Factors versus Manifest Nodes

For overall interpretation purposes, looking at factor-level drivers can be illuminating. Often, it provides a useful big-picture view of the domain. In order to identify specific product actions, however, we need to consider the manifest-level driver. As pointed out earlier, the factor-level drivers only exist as theoretical constructs, which cannot be directly observed in data. As a result, changing the factor nodes requires the manipulation of the underlying manifest nodes. For this reason, we now switch back our view of the network in order to only consider the manifest nodes in the analysis. We do that by right-clicking the **Use-of-Classes** icon ( ) in the bottom right corner of the **Graph Panel** window. This brings up the list of all **Classes**, of which we only check the class *Manifest* (Figure 8.80). Now all factors are translucent and excluded from analysis.

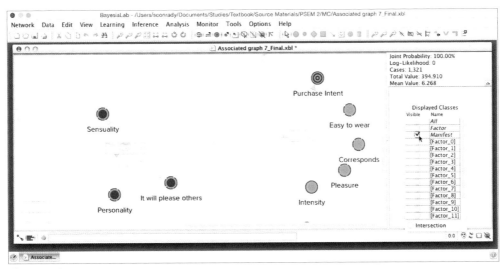

Figure 8.80

We repeat both the **Target Mean Analysis** (Figure 8.81) and the **Total Effects on Target** report.

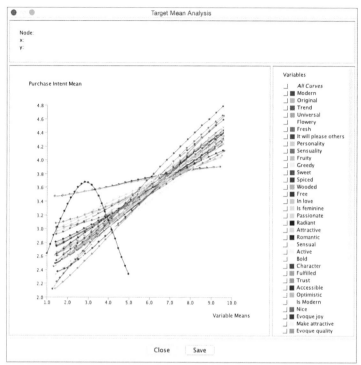

Figure 8.81

Not surprisingly, the **Manifest Nodes** show a similar pattern of association as the **Factors**. However, there is one important exception: the **Manifest Node** *Intensity* shows a nonlinear relationship with *Purchase Intent*. The curve for *Intensity* is shown with a gray line in Figure 8.81. Note that by hovering over a curve or a node name, BayesiaLab highlights the corresponding item in the legend or the plot respectively.

Also, we can see that *Intensity* was recorded on a 1–5 scale, rather than the 1–10 scale that applies to the other nodes. *Intensity* is a so-called "JAR" variable, i.e. a variable that has a "just-about-right" value. In the context of perfumes, this characteristic is obvious. A fragrance that is either too strong or too light is undesirable. Rather, there is a value somewhere in-between that would make a fragrance most attractive. The JAR characteristic is prototypical for variables representing sensory dimensions, e.g. saltiness or sweetness.

This emphasizes the importance of the visual analysis, as the nonlinearity goes unnoticed in the **Total Effects on Target** report (Figure 8.82). In fact, it drops almost to the bottom of the list in the report.

It turns out to be rather difficult to optimize a JAR-type variable at a population level. For example, increasing *Intensity* would reduce the number of consumers who found the fragrance too subtle. On the other hand, an increase in *Intensity* would presumably dismay some consumers who believed the original *Intensity* level to be appropriate.

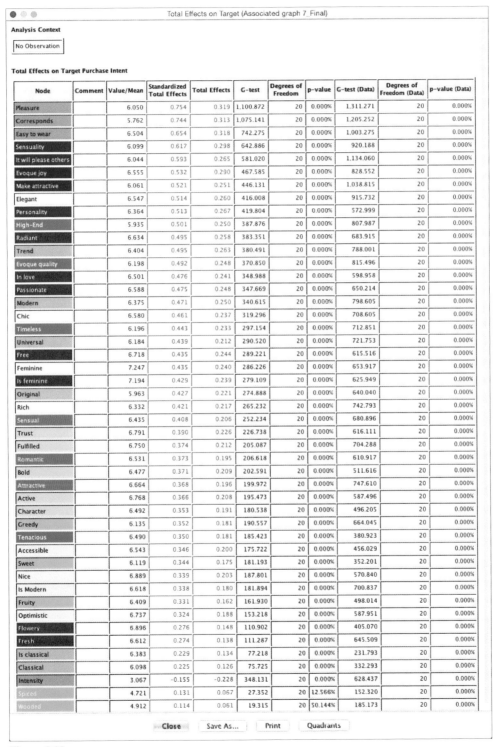

Figure 8.82

Constraints via Costs

As this drivers analysis model is intended to be used for product optimization, we need to consider any possible real-world constraints that may limit our ability to optimize any of the drivers in this domain. For instance, a perfumer may know how to change the intensity of a perfume but may not know how to directly affect the perception of "pleasure." In the original study, a number of such constraints were given.

In BayesiaLab, we can conveniently encode constraints via **Costs**, which is a **Node Property**. More specifically, we can declare any node as *Not Observable*, which—in this context—means that they cannot be considered with regard to optimization. Costs can be set by right-clicking on an individual node and then selecting **Properties > Cost** (Figure 8.83).

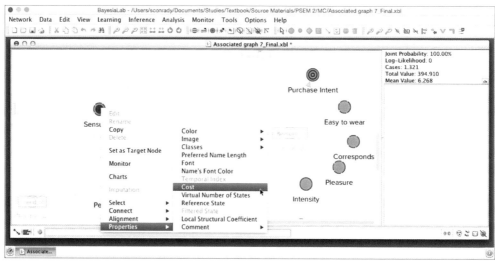

Figure 8.83

This brings up the **Cost Editor** for an individual node. By default, all nodes have a cost of 1.

Figure 8.84

Unchecking the box, or setting a value ≤0, results in the node becoming *Not Observable* (Figure 8.85).

Figure 8.85

Alternatively, we can bring up the **Cost Editor** for all nodes by right-clicking on the **Graph Panel** and then selecting **Edit Costs** from the contextual menu (Figure 8.86).

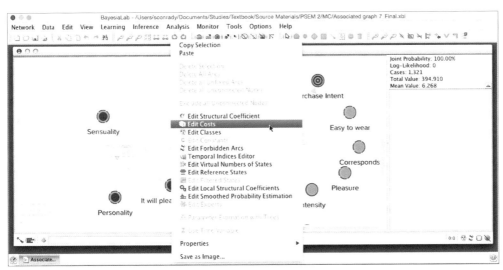

Figure 8.86

The **Cost Editor** presents the default values for all nodes (Figure 8.87).

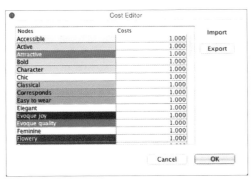

Figure 8.87

Again, setting values to zero will make nodes *Not Observable*. Instead of applying this setting node by node, we can import a cost dictionary that defines the values for each node. An excerpt from the text file is shown in Figure 8.88. The syntax is straightforward: *Not Observable* is represented by 0.

Figure 8.88

From within the **Cost Editor**, we can use the **Import** function to associate a cost dictionary. Alternatively, we can select **Data > Associate Dictionary > Node > Costs** from the main menu (Figure 8.89).

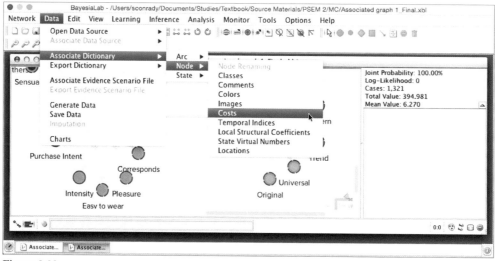

Figure 8.89

Upon import, the **Node Editor** reflects the new values (Figure 8.90), and the presence of non-default values for costs is indicated by the **Cost** icon ( ) in the lower right-hand corner of the **Graph Panel** window.

264

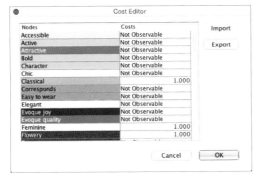

Figure 8.90

Furthermore, upon defining costs, we can see that all *Not Observable* nodes are marked with a pastel background (◯), as shown in Figure 8.91.

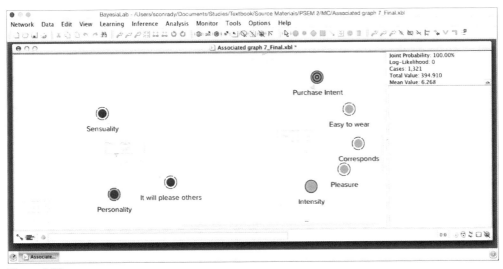

Figure 8.91

It is important to point out that all **Factors** are also set to *Not Observable* in our example. In fact, we do have two options here:

1. The optimization can be done at the first level of the hierarchical model, i.e. using the manifest variables;
2. The optimization can be performed at the second level of the model, i.e. using the **Factors**.

Most importantly, these two approaches cannot be combined as setting evidence on **Factors** will block information coming from manifest variables. Formally declaring the **Factors** as *Not Observable* tells BayesiaLab to proceed with option 1. Indeed, our plan is to perform optimization using the manifest variables only.

Multi-Quadrant Analysis

The network we have analyzed thus far modeled *Purchase Intent* as a function of perceived perfume characteristics. It is important to point out that this model represents the entire domain of all 11 tested perfumes. It is reasonable to speculate, however, that different perfumes have different drivers of *Purchase Intent*. Furthermore, for purposes of product optimization, we certainly need to look at the dynamics of each product individually.

BayesiaLab assists us in this task by means of **Multi-Quadrant Analysis**. This is a function that can generate new networks as a function of a breakout node in an existing network. This is the point where the node *Product* comes into play, which has been excluded all this time. Our objective is to generate a set of networks that model the drivers *Purchase Intent* for each perfume individually, as identified by the *Product* breakout variable.

We start the **Multi-Quadrant Analysis** by selecting **Tools > Multi-Quadrant Analysis** (Figure 8.92).

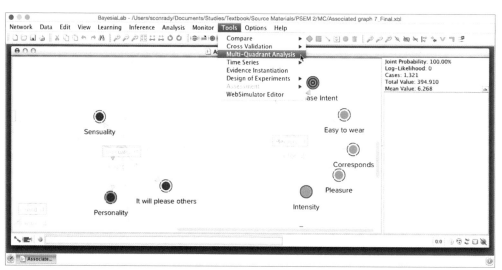

Figure 8.92

This brings up the dialog box, in which we need to set a number of options (Figure 8.93):

Chapter 8

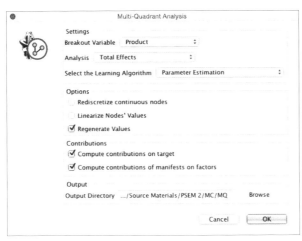

Figure 8.93

Firstly, **Breakout Variable** must be set to *Product* to indicate that we want to generate a network for each state of *Product*. For **Analysis**, we have a several options: We choose **Total Effects** to be consistent with the earlier analysis. Regarding the **Learning Algorithm**, we select **Parameter Estimation**. This choice becomes obvious once the database representing the "overall market" is split into 11 product-specific subsets. Now, the number of available observations per product drops to only 120. Given that most of our variables have 5 states, learning a structure with a database that small would be challenging.

▶ Discretization Intervals in Chapter 6, p. 119.

This also explains why we used the entire dataset to learn the PSEM structure, which will be shared by all the products. However, using **Parameter Estimation** will ensure that the parameters, i.e. the probability tables of each network, will be estimated based on the subsets of database records associated with each state of *Product*.

Among the **Options**, we check **Regenerate Values**. This recomputes, for each new network, the values associated with each state of the discretized nodes based on the respective subset of data.

There is no need to check **Rediscretize Continuous Nodes** because all discretized nodes share the same variation domain, and we require equal distance binning. However, we do recommend to use this option if the variation domains are different between subsets in a study, e.g. sales volume in California versus Vermont. Without using the **Rediscretize Continuous Nodes** option, it could happen that all data points for sales in Vermont end up in the first bin, effectively transforming the variable into a constant.

Furthermore, we do not check the option for **Linearize Nodes' Values** either. This function reorders a node's states so that its states' values have a monotonically positive relationship with the values of the **Target Node**. Applying this transformation to the node *Intensity* would artificially increase its impact. It would incorrectly imply that it is possible to change a perfume in a way that simultaneously satisfies those consumers who rated it as too subtle and also those who rated it as too strong. Needless to say, this is impossible.

Finally, computing all **Contributions** will be helpful for interpreting each product-specific network.

Upon clicking **OK**, 11 networks are created and saved to the output directory defined in the dialog box. Each network is then analyzed with the specified **Analysis** method to produce the **Multi-Quadrant Plot**. Figure 8.94 shows the **Quadrant Plot** for *Product 1*.

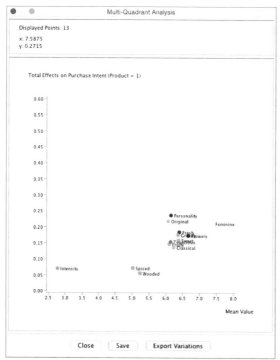

Figure 8.94

The x-value of each point indicates the mean value of the corresponding manifest variable, as rated by those respondents who have evaluated *Product 1*; the position on the y-axis reflects the computed **Total Effect**.

From the contextual menu (Figure 8.95), we can choose **Display Horizontal Scales** and **Display Vertical Scales**, which provides the range of positions of the other products.

Chapter 8

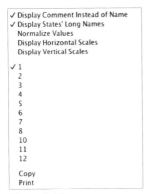

Figure 8.95

Using **Horizontal Scales** provides a quick overview of how the product under study is rated vis-à-vis other products (Figure 8.96). The **Vertical Scales** compare the importance of each dimension with respect to *Purchase Intent*. Finally, we can select the individual product to be displayed in the **Multi-Quadrant Analysis** window via the **Contextual Menu**.

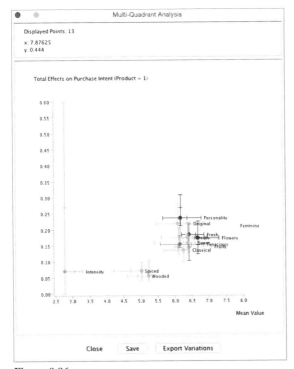

Figure 8.96

Drawing a rectangle with the cursor zooms in on the specified area of the plot (Figure 8.97).

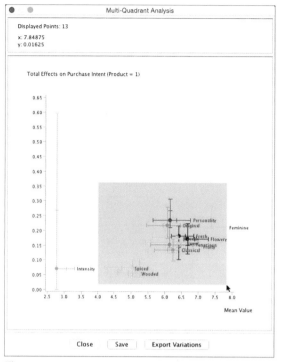

Figure 8.97

The meaning of the **Horizontal Scales** and **Vertical Scales** becomes apparent when hovering over any dot as this brings up the position of the other (competitive) products with regard to the currently highlighted attribute (Figure 8.98).

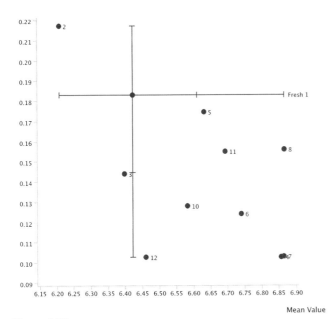

Figure 8.98

In Figure 8.98, this means, for instance, that *Product 2* and *Product 7* are rated lowest and highest respectively on the x-scale with regard to the variable *Fresh*. In terms of **Total Effect** on *Purchase Intent*, *Product 12* and *Product 2* mark the bottom and top end respectively (y-scale).

From a product management perspective, this suggests that for *Product 1*, with regard to the attribute *Fresh*, there is a large gap to the level of the best product, i.e. *Product 7*. So, one could interpret the variation from the status quo to the best level as "room for improvement" for *Product 1*.

On the other hand, as we can see in Figure 8.99 (scales omitted), the variables *Personality, Original* and *Feminine*, and have a greater **Total Effect** on *Purchase Intent*. These relative positions will soon become relevant as we will need to simultaneously consider improvement potential and importance for optimizing *Purchase Intent*.

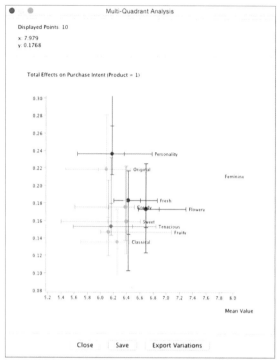

Figure 8.99

BayesiaLab's **Export Variations** function allows us to save the variation domain for each driver, i.e. the minimum and maximum mean values observed across all products in the study.

Knowing these variations will be useful for generating realistic scenarios for the subsequent optimization. However, what do we mean by "realistic"? Ultimately, only a subject matter expert can judge how realistic a scenario is. However, a good heuristic is whether or not a certain level is achieved by any product in the market. One could argue that the existence of a certain satisfaction level for some product means that such a level is not impossible to achieve and is, therefore, "realistic."

Clicking the **Export Variations** button (Figure 8.100) saves the absolute variations to a text file (Figure 8.101) for subsequent use in optimization.

Figure 8.100

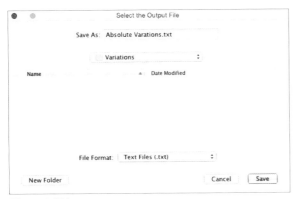

Figure 8.101

Absolute Varations.txt

```
#Fri Jul 10 09:33:54 CDT 2015
Fresh=6.208333492279053 6.866666793823242
Flowery=6.583333492279053 7.305785179138184
Greedy=5.610267639160156 6.572437286376953
Spiced=4.133597373962402 5.063426971435547
Wooded=4.578180313110352 5.261065483093262
Sweet=5.400458335876465 6.6534833908081055
Fruity=6.002670764923096 7.091536521911621
Original=5.4666666984558105 6.4666666984558105
Feminine=6.800000190734863 7.8416666984558105
Tenacious=5.453145503997803 6.840022563934326
Personality=5.680275917053223 6.820230484008789
Classical=5.709170341491699 6.449088096618652
Intensity=2.7750000953674316 3.299999952316284
```

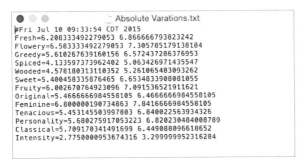

Figure 8.102

Product Optimization

In order to perform optimization for a particular product, we need to open the network for that specific product. Networks for all products were automatically generated and saved during the **Multi-Quadrant Analysis** (Figure 8.103), so we simply need to open the network for the product of interest. The suffix in the file name reflects the *Product*.

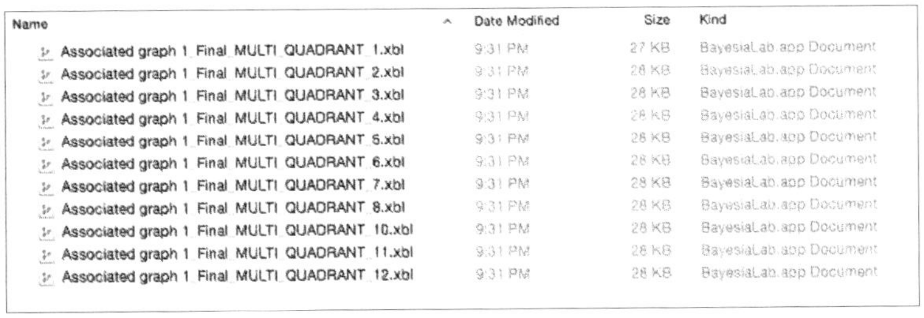

Figure 8.103

To demonstrate the optimization process, we open the file that corresponds to *Product 1* (Figure 8.104). Structurally, this network is identical to the network learned from the entire dataset. However, the parameters of this network were estimated only on the basis of the observations associated with *Product 1*.

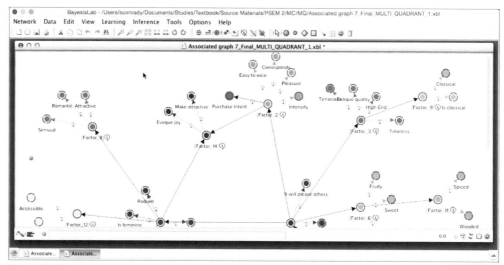

Figure 8.104

Now we have all the elements that are necessary for optimizing the *Purchase Intent* of *Product 1*:

- A network specific to *Product 1*;
- A set of driver variables, selected by excluding the non-driver variables via **Costs**;
- Realistic scenarios, as determined by the **Variation Domains** of each driver variable.

With the above, we are now in a position to search for node values that optimize *Purchase Intent*.

Target Dynamic Profile

Before we proceed, we need to explain what we mean with optimization. As all observations in this study are consumer perceptions, it is clear that we cannot directly manipulate them directly. Rather, the purpose of this optimization is to identify in which order these perceptions should be addressed by the perfume maker. Some consumer perceptions may relate to specific odoriferous compounds that a perfumer can modify; other perceptions can perhaps be influenced by marketing and branding initiatives. However, the precise mechanism of influencing consumer perceptions is

Chapter 8

not the subject of our discussion. From our perspective, the causes that could influence the perception are hidden. Thus, we have here a prototypical application of **Soft Evidence**, i.e. we assume that the simulated changes in the distribution of consumer perceptions originate in hidden causes.

▶ Numerical Evidence in Chapter 7, p. 190.

While BayesiaLab offers a number of optimization methods, **Target Dynamic Profile** is appropriate here. We start it from within **Validation Mode** (≡ or F5) by selecting **Analysis > Report > Target Analysis > Target Dynamic Profile** (Figure 8.105).

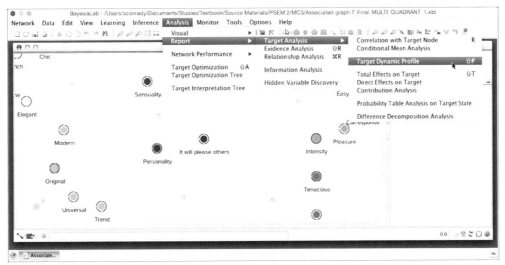

Figure 8.105

We need to explain the large number of options that must be set for **Target Dynamic Profile** (Figure 8.106). These options will reflect our objective of pursuing realistic sets of evidence:

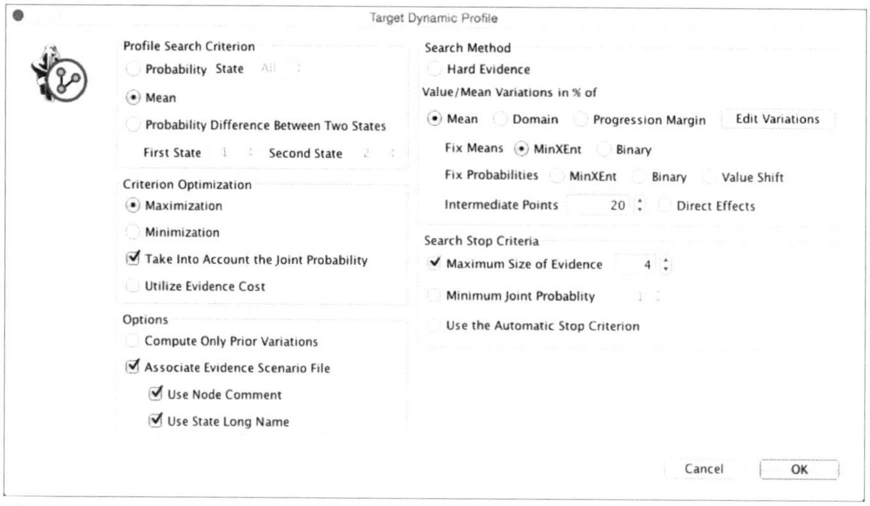

Figure 8.106

In **Profile Search Criterion** we specify that we want to optimize the mean value of the **Target Node**, as opposed to any particular state or the difference between states.

Joint Probability

Next, we specify under **Criterion Optimization** that the mean value of the **Target Node** is to be maximized. Furthermore, we check **Take Into Account the Joint Probability**. This weights any potential improvement in the mean value of **Target Node** by the joint probability that corresponds to the set of simulated evidence that generated this improvement. The joint probability of a simulated evidence scenario will be high if its probability distribution is close to the original probability distribution observed in the consumer population: the higher the joint probability, the closer is the simulated scenario to the status quo of customer perception.

In practice, checking this option means that we prefer smaller improvements with a high joint probability over larger ones with a low joint probability. Figure 8.107 illustrates this point: $0.146 \times 26.9\% = 0.0393 > 0.174 \times 21.33\% = 0.0371$.

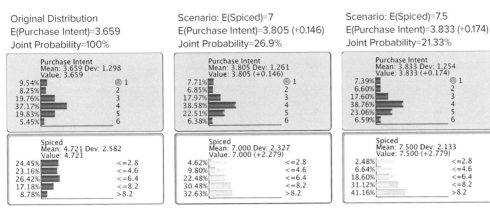

Figure 8.107

If all simulated values were within the constraints set in the **Variation Editor**, it would be better to increase the driver variable *Spiced* to a simulated value of 7 rather than 7.5, even though *Purchase Intent* would be higher for the latter value of *Spiced*. In other words, the "support" for E(*Spiced*)=7 is greater than for E(7.5), as more respondents are already in agreement with such a scenario. Once again, this is about pursuing improvements that are achievable rather than proposing pie-in-the-sky scenarios.

Costs

In this example, so far, we have only used **Costs** for selecting the subset of driver variables. Additionally, we can utilize **Costs** in the original sense of the word in the optimization process. For instance, if we had information on the typical cost of improving a specific rating by one unit, we could enter such a value as cost. This could be a dollar value, or we could set the costs in such a way that they reflect the relative effort required for the same amount of change, e.g. one unit, in each of the driver variables. For example, a marketing manager may know that it requires twice as much effort to change the perception of *Feminine* compared to changing the perception of *Sweet*. If we want to quantify such efforts by using **Costs**, we will need to ensure that the costs of all variables share the same scale. For instance, if some drivers are measured in dollars, and others are measured in terms of time spent in hours, we will need to convert hours to dollars.

In our study, we leave all the included driver variables at a cost of 1, i.e. we assume that it requires the same effort for the same amount of change in any driver variable. Hence, we can leave the **Utilize Evidence Cost** unchecked.[10]

---

10   **Not Observable** nodes still remain excluded as driver variables.

Compute Only Prior Variations needs to remain unchecked as well. This option would be useful if we were interested in only computing the marginal effect of drivers. For that purpose, we would not want any cumulative effects or conditional variations given other drivers.

Associate Evidence Scenario will save the identified sets of evidence for subsequent evaluation.

The setting **Search Methods** is critically important for the optimization task. We need to define how to search for sets of evidence. Using **Hard Evidence** means that would we exclusively try out sets of evidence consisting of nodes with one state set to 100%. This would imply that we simulate a condition in which all consumers perfectly agree with regard to some ratings. Needless to say, this would be utterly unrealistic. Instead, we will explore sets of evidence, consisting of distributions for each node, by modifying their mean values as **Soft Evidence**. More precisely, we use the **MinXEnt** method to generate such evidence.

▶ MinXEnt ("Minimum Cross-Entropy") in Chapter 7, p. 191.

In this context, we reintroduce the variations we saved earlier. We reason that the best-rated product with regard to a particular attribute represents a plausible upper limit for what any product could strive for in terms of improvement. This also means that a driver variable that has already achieved the best level will not be optimized any further in this framework.

Variation Editor

Clicking on **Variations** brings up the Variation Editor (Figure 8.108). By default, it shows variations in the amount of ±100% of the current mean.

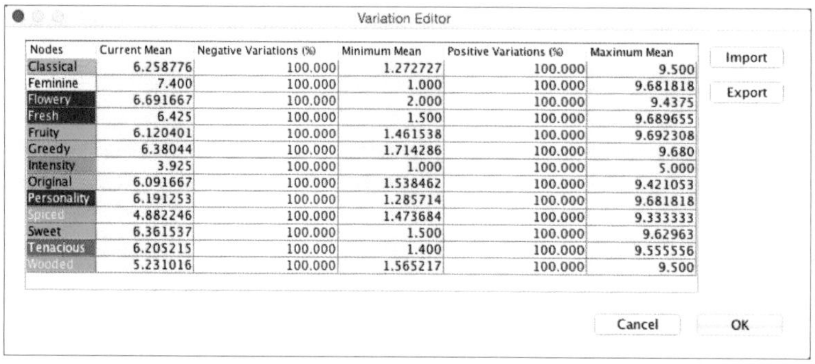

Nodes	Current Mean	Negative Variations (%)	Minimum Mean	Positive Variations (%)	Maximum Mean
Classical	6.258776	100.000	1.272727	100.000	9.500
Feminine	7.400	100.000	1.000	100.000	9.681818
Flowery	6.691667	100.000	2.000	100.000	9.4375
Fresh	6.425	100.000	1.500	100.000	9.689655
Fruity	6.120401	100.000	1.461538	100.000	9.692308
Greedy	6.38044	100.000	1.714286	100.000	9.680
Intensity	3.925	100.000	1.000	100.000	5.000
Original	6.091667	100.000	1.538462	100.000	9.421053
Personality	6.191253	100.000	1.285714	100.000	9.681818
Spiced	4.882246	100.000	1.473684	100.000	9.333333
Sweet	6.361537	100.000	1.500	100.000	9.62963
Tenacious	6.205215	100.000	1.400	100.000	9.555556
Wooded	5.231016	100.000	1.565217	100.000	9.500

Figure 8.108

To load the **Variations** that we generated through **Multi-Quadrant Analysis**, we click **Import** and select **Absolute Variations** from the pop-up window.

Figure 8.109

Now we can open the previously saved file.

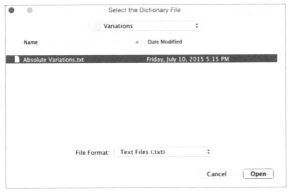

Figure 8.110

The **Variation Editor** now reflects the constraints. Any available expert knowledge can be applied here, either by entering new values for the **Minimum Mean** or **Maximum Mean**, or by entering percent values for **Positive Variations** and **Negative Variations**.

Depending on the setting, the percentages are relative to (a) the **Current Mean**, (b) the **Domain**, or (c) the **Progression Margin** as illustrated in Figure 8.111.

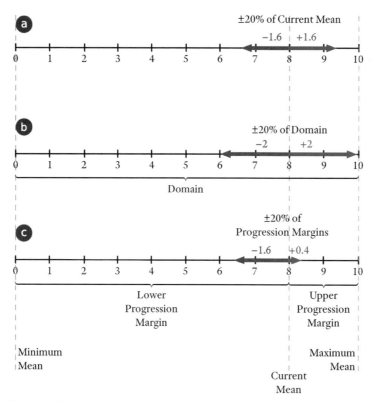

Figure 8.111

Using the **Progression Margin** is particularly useful as it automatically constrains the positive and negative variations in proportion to the gap from the current mean to the maximum and minimum mean values respectively. In other words, it limits the improvement potential of a driver variable as its value approaches the maximum. It is a practical—albeit arbitrary—approach to prevent overly optimistic optimizations.

Nodes	Current Mean	Negative Variations (%)	Minimum Mean	Positive Variations (%)	Maximum Mean
Classical	6.258333	8.921441	5.700	2.929426	6.441667
Feminine	7.400	8.108106	6.800	5.968469	7.841667
Flowery	6.691667	1.618927	6.583333	9.177363	7.305785
Fresh	6.425	3.372241	6.208333	6.874191	6.866667
Fruity	6.125	1.904764	6.008333	15.646261	7.083333
Greedy	6.383333	12.010442	5.616667	2.088776	6.516667
Intensity	2.775	0.000	2.775	18.918917	3.300
Original	6.091667	10.259917	5.466667	6.155951	6.466667
Personality	6.183333	8.625335	5.650	9.703504	6.783333
Spiced	5.041667	16.363636	4.216667	0.000	5.041667
Sweet	6.400	15.494792	5.408333	3.255205	6.608333
Tenacious	6.166667	9.324324	5.591667	11.101184	6.85124
Wooded	5.241667	11.923686	4.616667	0.000002	5.241667

Figure 8.112

Next, we select **MinXEnt** in the **Search Method** panel as the method for generating **Soft Evidence**. In terms of **Intermediate Points**, we set a value of 20. This means that

BayesiaLab will simulate 22 values for each node, i.e. the minimum and maximum plus 20 intermediate values, all within the constraints set by the variations. This is particularly useful in the presence of non-linear effects.

Within the **Search Stop Criteria** panel, **Maximum Size of Evidence** specifies the maximum number of driver variables to be recommended as part of the optimization policy. This setting is once again driven by real-world considerations. Although one could wish to bring all variables to their ideal level, a decision maker may recognize that it is not plausible to pursue anything beyond the top-4 variables.

Alternatively, we can choose to terminate the optimization process once the joint probability of the simulated evidence drops below the specified **Minimum Joint Probability**.

The final option, **Use the Automatic Stop Criterion**, leaves it up BayesiaLab to determine whether adding further evidence provides a worthwhile improvement for the **Target Node**.

Optimization Results

Once the optimization process concludes, we obtain a report window that contains a list of priorities: *Personality, Fruity, Flowery,* and *Tenacious* (Figure 8.113).

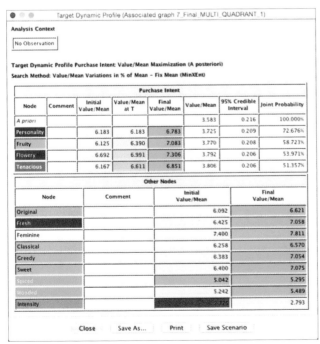

Figure 8.113

To explain the items in the report, we present a simplified and annotated version in Figure 8.114. Note that this report can be saved in HTML format, for subsequent editing as a spreadsheet. Preparing Figure 8.114 is an example of that approach.

Analysis Context	
No Observation	No other evidence is set

Search Method: Value/Mean Variations in % of Mean - Fix Mean (MinXEnt)					
Target Dynamic Profile Purchase Intent: Value/Mean Maximization (A posteriori)					
Purchase Intent					
Node	Initial Value/Mean	Value/Mean at T	Final Value/Mean	Value/Mean	Joint Probability
				Initial value of Purchase Intent, prior to optimization. ▼	Initial joint probability is 100%. ▼
			A priori ▶	3.583	100.00%
Personality	6.183	6.183	6.783	3.725	72.68%
▲ Most important node	▲ Initial value of *Intensity*, prior to optimization		▲ Optimal value of *Intensity*, within given constraints.	▲ Value of *Purchase Intent* after *Personality* is set to optimal value.	▲ New joint probability after setting *Personality*. This means that 72.68% of the observations already meet this condition.
				This means that after setting *Personality* to the optimal value, *Purchase Intent* increases from 3.583 to 3.725.	
Fruity	6.125	6.39	7.083	3.77	58.72%
▲ 2nd most important node	▲ Initial value of *Fruity*, prior to optimization.	▲ Value of *Fruity*, after *Personality* is set to optimal value.	▲ Optimal value of *Fruity*, within given constraints.	▲ Value of *Purchase Intent*, after *Personality* and *Fruity* are set to optimal values.	▲ New joint probability after setting *Personality*.
Flowery	6.692	6.991	7.306	3.792	53.97%
▲ 3rd most important node	▲ Initial value of *Flowery*, prior to optimization.	▲ Value of *Flowery*, after *Personality* and *Fruity* are set to optimal value.	▲ Optimal value of *Flowery*, within given constraints.	▲ Value of *Purchase Intent*, after *Personality*, *Fruity*, and *Flowery* are set to optimal values.	▲ New joint probability after setting *Flowery*.
Tenacious	6.167	6.611	6.851	3.806	51.36%
▲ 4th most important node	▲ Initial value of *Tenacious*, prior to optimization.	▲ Value of *Tenacious*, after *Personality*, *Fruity*, and *Flowery* are set to optimal values.	▲ Optimal value of *Tenacious*, within given constraints.	▲ Value of *Purchase Intent*, after *Personality*, *Fruity*, *Flowery*, and *Tenacious* are set to optimal values.	▲ New joint probability after setting *Tenacious*.
				▲ This means that after applying all four listed measures, an increase of 0.223 would be observed for *Purchase Intent*.	

Other Nodes		
Node	Initial Value/Mean	Final Value/Mean
Original	6.092	6.621
	▲ Initial value of *Original*, prior to optimization.	▲ Final value of *Original*, after setting the top-4 nodes to the optimal levels.
Fresh	6.425	7.058
	▲ Initial value of *Fresh*, prior to optimization.	▲ Final value of *Fresh*, after setting the top-4 nodes to the optimal levels.
Feminine	7.4	7.811
	▲ Initial value of *Feminine* prior to optimization.	▲ Final value of *Feminine*, after setting the top-4 nodes to the optimal levels.

Collateral Effects: Even though *Other Nodes* do not directly belong to the recommended group of top drivers, the predicted level of Purchase Intent can only be reached with certainy if the *Other Nodes* can be set to these levels.

Figure 8.114

Most importantly, the **Value/Mean** column shows the successive improvement upon implementation of each policy. From initially 3.58, the *Purchase Intent* improves to

3.86, which may seem like a fairly small step. However, the importance lies in the fact that this improvement is not based on Utopian thinking, but rather on modest changes in consumer perception, well within the range of competitive performance.

Evidence Scenarios

As an alternative to interpreting the static report, we can examine each element in the list of priorities. To do so, we bring up all the **Monitors** of the nodes identified for optimization.

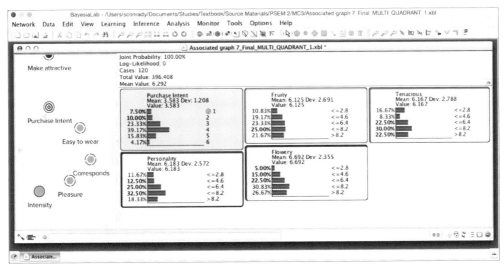

Figure 8.115

Then, we retrieve the individual steps by right-clicking on the **Evidence Scenario** icon ( ) in the lower right-hand corner of the main window (Figure 8.116).

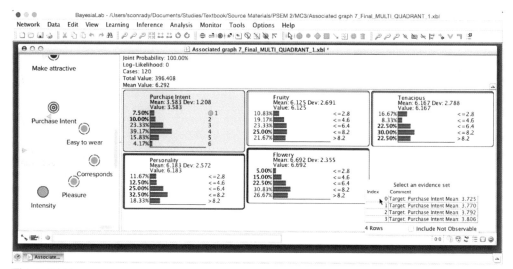

Figure 8.116

Selecting the first row in the table (Index=0) sets the evidence that corresponds to the first priority (Figure 8.116), i.e. *Personality*. We can now see that the evidence we have set is a distribution, rather than a single value. The small gray arrows indicate how the distribution of the evidence and the distributions of *Purchase Intent, Fruity, Flowery,* and *Tenacious* are all different from their prior, marginal distributions (Figure 8.117). The changes to the *Fruity, Flowery,* and *Tenacious* correspond what is shown in the report in the column **Value/Mean at T** (Figure 8.114).

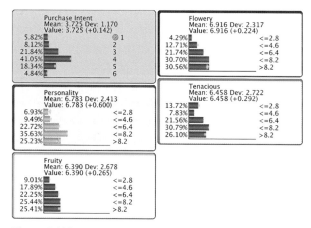

Figure 8.117

By selecting Index=1 we introduce a second set of evidence, i.e. the optimized distribution for *Personality* (Figure 8.118).

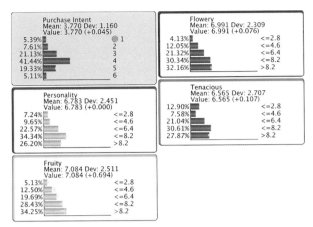

Figure 8.118

Continuing with Index 2 and 3, we see that the improvements to *Purchase Intent* become smaller (Figure 8.119 and Figure 8.120).

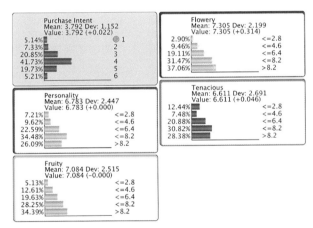

Figure 8.119

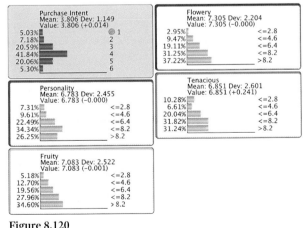

Figure 8.120

Bringing up all the remaining nodes would bring up any "collateral" changes as a result of setting multiple pieces of evidence.

The results tell us that for *Product 1*, a higher consumer rating of *Personality* would be associated with a higher *Purchase Intent*. Improving the perception of *Personality* might be a task for the marketing and advertising team. Similarly, a better consumer rating of *Fruity* would also be associated with greater *Purchase Intent*. A product manager could then interpret this and request a change to some ingredients. Our model tells us that, if such changes in consumer ratings were to be brought about in the proposed order, a higher *Purchase Intent* would be potentially be observed.

While we have only presented the results for *Product 1*, we want to highlight that the priorities are indeed different for each product, even though they all share the same underlying PSEM structure. The recommendations from the **Target Dynamic Profile** of *Product 11* are shown in Figure 8.121.

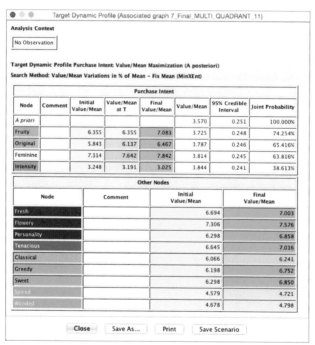

Figure 8.121

This is an interesting example as it identifies that the JAR-type variable *Intensity* needs to be lowered to optimize *Purchase Intent* for *Product 11*.

It is important to reiterate that the sets of evidence we apply are not direct interventions in this domain. Hence, we are not performing causal inference. Rather, the sets of evidence we found help us prioritize our course of action for product and marketing initiatives.

Summary

We presented a complete workflow that generates a Probabilistic Structural Equation Model for key drivers analysis and product optimization. The Bayesian networks paradigm turned out to a practical platform for the development of the model and its subsequent analysis, all the way through optimization. With all steps contained in BayesiaLab, we have a single, continuous line of reasoning from raw survey data to a final order of priorities for action.

Chapter 9

9. Missing Values Processing

Missing values are encountered in virtually all real-world data collection processes. Missing values can be the result of non-responses in surveys, poor record-keeping, server outages, attrition in longitudinal surveys, or the faulty sensors of a measuring device, etc. Despite the intuitive nature of this problem, and the fact that almost all quantitative studies are affected by it, applied researchers have given it remarkably little attention in practice. Burton and Altman (2004) state this predicament very forcefully in the context of cancer research: "We are concerned that very few authors have considered the impact of missing covariate data; it seems that missing data is generally either not recognized as an issue or considered a nuisance that it is best hidden."

Given the abundance of "big data" in the field of analytics, missing values processing may not be a particularly fashionable topic. After all, who cares about few missing data points if there are many more terabytes of observations waiting to be processed? One could be tempted to analyze complete data only and simply remove all incomplete observations. Regardless of how many more complete observations might be available, this naive approach would almost certainly lead to misleading interpretations or create a false sense of confidence in one's findings.

Koller and Friedman (2009) provide an example of a hypothetical medical trial that evaluates the efficacy of a drug. In this trial, patients can drop out, in which case their results are not recorded. If patients withdraw at random, there is no problem ignoring the corresponding observations. On the other hand, if patients prematurely quit the trial because the drug does not seem to help them, discarding these observations introduces a strong bias in the efficacy evaluation. As this example illustrates, it is important to understand the mechanism that produces the missingness, i.e. the conditions under which some values are not observed.

As missing values processing—beyond the naive ad hoc approaches—can be a demanding task, both methodologically and computationally, the objective of this chapter is to demonstrate how advanced missing values processing methods can be integrated into a research workflow with BayesiaLab.

We have already mentioned missing values processing several times in earlier chapters, as it is one of the steps in the **Data Import Wizard**. However, we have delayed a formal discussion of the topic until now because the recommended missing values processing methods are tightly integrated with BayesiaLab's learning algorithms. Indeed, all of BayesiaLab's core functions for learning and inference are prerequisites for the successful application of missing values processing. With all building blocks in place, we can now explore this subject in detail.

Types of Missingness

There are four principal types of missing values that are typically encountered in research:

1. Missing Completely at Random (MCAR)
2. Missing at Random (MAR)
3. Missing Not at Random (MNAR) or Not Missing at Random (NMAR)[1]
4. Filtered Values

We will now exemplify each of these conditions with a causal Bayesian network. In this format, we can conveniently illustrate (a) the data-generating process (DGP), (b) the mechanism that causes the missingness, and (c) the observable variables that contain the missing values (Figure 9.1).

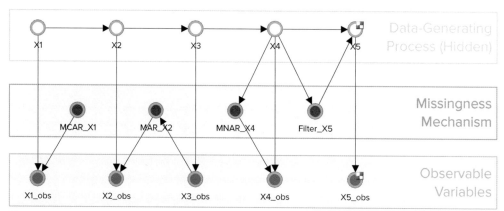

Figure 9.1

Furthermore, we can use this causal network directly to simulate all missingness conditions and evaluate their (mostly adverse) effects. In the second half of this chapter, we take this experiment a step further by generating sample data from this model.

---

1 Both of these equivalent expressions, MNAR and NMAR, appear equally frequently in the literature. We use MNAR throughout this chapter.

Chapter 9

With actual data, the problems associated with missingness become tangible. Beyond highlighting these challenges, we will put BayesiaLab to the test by feeding it the incomplete data in an attempt to recover the original distributions. As it turns out, Bayesian networks provide a perfect platform for overcoming the complications caused by missing data.

Missing Completely at Random

Missing Completely at Random (MCAR) means that the missingness mechanism is entirely independent of all other variables. In our causal Bayesian network, we encode this independent mechanism with a boolean variable named *MCAR_X1*.

Furthermore, we assume that there is a variable *X1* that represents the original data-generating process. This variable, however, is hidden, so we cannot observe it directly. Rather, we can only observe *X1* via the variable *X1_obs*, which is a "clone" of *X1* but with one additional state, "?", which indicates that the value of *X1* is not observed.

The Bayesian network shown in Figure 9.2 is a subnetwork of the complete network of Figure 9.1. The behavior of the three variables we just described is encoded in this subnetwork.

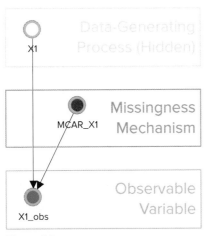

Figure 9.2

In addition to this qualitative structure, we need describe the quantitative part, i.e. the parameters of this subnetwork, including the missingness mechanism and the observable variable:

- *X1* is a continuous variable with values between 0 and 1. We have arbitrarily defined a Normal distribution for modeling the DGP.
- *MCAR_X1* is a boolean variable without any parent nodes. This means that *MCAR_X1* is independent of all variables, whether hidden or not. Its probability of being *true* is 10%.
- *X1_obs* has two parents: the data-generating variable *X1* and the missingness mechanism *MCAR_X1*. The conditional probability distribution of *X1_obs* is defined by the following deterministic rule:

$$\text{IF} \quad MCAR_{X1} \quad \text{THEN} \quad X1_{obs} = ? \quad \text{ELSE} \quad X1_{obs} = X1 \tag{9.1}$$

Now that our causal Bayesian network is fully specified, we can evaluate the impact of the missingness mechanism on the observable variable *X1_obs*. Given that we have created a complete model of this small domain, we automatically have perfect knowledge of the distribution of *X1*. Thus, we can directly compare *X1* and *X1_obs* via the **Monitors** (Figure 9.3).

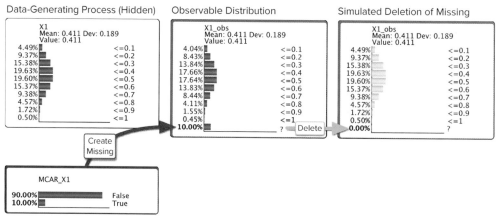

Figure 9.3

We see that *X1* (left) and *X1_obs* (center) have the same mean and the same standard deviation. This suggests that the remaining observations in *X1_obs* (center) are not different from the non-missing cases in *X1* (left). The only difference is that *X1_obs* (center) has one additional state ("?") for missing values, representing 10% of the observations. Thus, deleting the missing observations of an MCAR variable should not bias the estimation of its distribution. In BayesiaLab, we can simulate this assumption by setting negative evidence on "?" (green arrow labeled "Delete"). As we can see, the distribution of *X1_obs* (right) is now exactly the same as the one of *X1* (left).

Under real-world conditions, however, we typically do not know whether the missing values in our dataset were generated completely at random (MCAR). This would be a strong assumption to make, and it is generally not testable. As a result, we can rarely rely on this fairly benign condition of missingness and, thus, should never be too confident in deleting missing observations.

Missing at Random

Secondly, data can be Missing at Random (MAR). Here, the missingness of data depends on observed variables. A brief narrative shall provide some intuition for the MAR condition: in a national survey of small business owners about business climate, there is a question about the local cost of energy. Chances are that the owner of a business that uses little electricity, e.g. a yoga studio, may not know of the current cost of 1 kWh of electric energy and could not answer that question, thus producing a missing value in the questionnaire. On the other hand, the owner of an energy-intensive business, e.g. an electroplating shop, would presumably be keenly aware of the electricity price and able to respond accordingly. In this story, the probability of non-response is presumably inversely proportional to the energy consumption of the business.

In the subnetwork shown of Figure 9.4, *X3_obs* is the observed variable that causes the missingness, e.g. the energy consumption in our story. *X2_obs* would be the stated price of energy, if known. *X2* would represent the actual price of energy in our narrative. Indeed, from the researcher's point of view, the actual cost of energy in each local market and for each electricity customer is hidden.

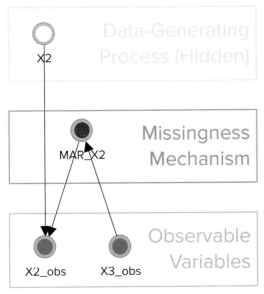

Figure 9.4

To simulate this network, we need to define its parameters, i.e. the quantitative part of the network structure:

- *X2* is a continuous variable with values between 0 and 1. Here, too, we have arbitrarily defined a Normal distribution for modeling the DGP.
- *MAR_X2* is a boolean variable with one parent, which specifies that the missingness probability depends directly on the fully observed variable *X3_obs*. The exact values are not important here, as we only need to know that the probabilities of missingness are inversely proportional to the values of *X3_obs*:

$$P(MAR_{X2} = true \mid X3_{obs}) \propto \frac{1}{X3_{obs}} \tag{9.2}$$

- *X2_obs* has two parents, i.e. the data generating variable *X2* and the missingness mechanism *MAR_X2*. The conditional probability distribution of *X2_obs* can be described by the following deterministic rule:

$$\texttt{IF} \quad MAR_{X2} \quad \texttt{THEN} \quad X2_{obs} = ? \quad \texttt{ELSE} \quad X2_{obs} = X2 \tag{9.3}$$

Given the fully specified network, we can now simulate the impact of the missingness mechanism on the observable variable *X2_obs*.

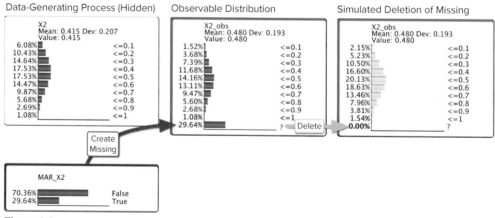

Figure 9.5

As Figure 9.5 shows, the mean and standard deviation in the **Monitor** of *X2_obs* indicates that the distribution of the observed values of *X2* differs significantly from the original distribution, leading to an overestimation of *X2* in this example. We can simulate the deletion of the incomplete observations by setting negative evidence on "?" in the **Monitor** of *X2_obs* (green arrow labeled "Delete"). The simulated distribution of *X2_obs* (right) clearly differs from the one of *X2* (left).

Chapter 9

Missing Not at Random

Missing Not at Random (MNAR) refers to situations in which the missingness of a variable depends on hidden causes (unobserved variables), such as the data-generating variable itself, for instance. This condition is depicted in the subnetwork of Figure 9.6.

An example of the MNAR condition would be a hypothetical telephone survey about alcohol consumption. Heavy drinkers might decline to provide an answer out of fear of embarrassment. On the other hand, survey participants who drink very little or nothing at all might readily report their actual drinking habits. As a result, the missingness is a function of the very variable in which we are interested.

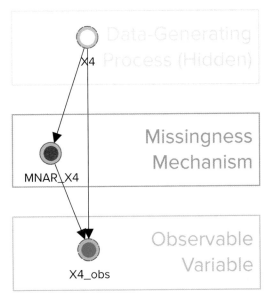

Figure 9.6

In order to proceed to simulation, we need to specify the parameters of the missingness mechanism and the observable variable:
- *X4* is a continuous variable with values between 0 and 1, and a Normal distribution models the DGP.
- *MNAR_X4* is a boolean variable with one parent, which specifies that the missingness probability depends directly on the hidden variable *X4*. However, the exact values are unimportant. We simply need to state that the probabilities of missingness are proportional to the values of *X4*:

$$P(MNAR_{X4} = true \mid X4) \propto X4 \qquad (9.4)$$

- *X4_obs* has two parents, i.e. the data-generating variable *X4* and the missingness mechanism *MNAR_X4*. The conditional probability distribution is defined by the following deterministic rule:

$$\text{IF } MNAR_{X4} \text{ THEN } X4_{obs} = ? \text{ ELSE } X4_{obs} = X4 \tag{9.5}$$

The impact of the missing values mechanism becomes apparent as we compare the **Monitors** of the network side by side (Figure 9.7).

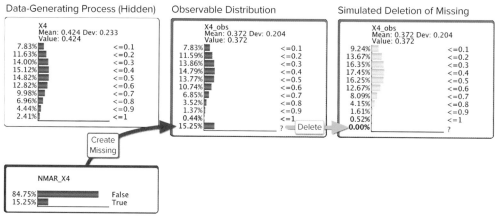

Figure 9.7

As Figure 9.7 shows, the mean and standard deviation in the **Monitor** of *X4_obs* (center column) indicate that the distribution of the observed values of *X4* differs significantly from the original distribution (left column), leading to an underestimation of *X4* in this example. We can simulate the deletion of the incomplete observations by setting negative evidence on "?" (green arrow labeled "Delete"). The simulated distribution of *X4_obs* (right column) indeed differs from the one of *X4* (left column).

Filtered Values

▸ Filtered Values in Chapter 5, p. 84.

There is a fourth type of missingness, which is less often mentioned in the literature. In BayesiaLab, we refer to missing data of this kind as **Filtered Values**. In fact, **Filtered Values** are technically not missing at all. Rather, **Filtered Values** are values that do not exist in the first place. Clearly, something nonexistent cannot become missing as a result of a missingness mechanism.

For instance, in a hotel guest survey, a question about one's satisfaction with the hotel swimming pool cannot be answered if the hotel property does not have a swimming pool. This question is simply not applicable. The absence of a swimming pool rating for this hotel *would not* be a missing value. On the other hand, for a hotel with a swimming pool, the absence of an observation *would* be a missing value.

Conceptually, **Filtered Values** are quite similar to MAR values, as **Filtered Values** usually depend on other variables in the dataset, too, which may or may not be fully observed. However, **Filtered Values** should never be processed as missing values. In our example, it is certainly not reasonable to impute a value for the swimming pool rating if there is no swimming pool. Rather, a **Filtered Value** should be considered a special type of observation.

In BayesiaLab, an additional state, marked with a chequered icon (▨), is added to this type of variable in order to denote **Filtered Values**.[2] Figure 9.8 shows an example of a network including **Filtered Values**.

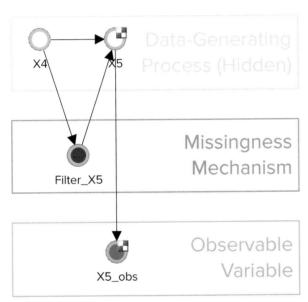

Figure 9.8

Once again, we must describe the parameters of the subnetwork, including the **Filtered Values** mechanism and the observable variable:

- *Filter_X5* is a boolean variable with one parent, which specifies that it depends on the hidden variable *X4*. Here, *X5* becomes a **Filtered Value** if *X4* is greater than 0.7.

$$Filter_{X5} = X4 > 0.7 \tag{9.6}$$

---

2 BayesiaLab's learning algorithms implement a kind of local selection for excluding the observations with **Filtered Values** while estimating the probabilistic relationships.

- *X5* is a continuous variable with values between 0 and 1. It has two parents, *X4* and the **Filtered Values** mechanism.

$$\texttt{IF} \quad Filter_{X5} \quad \texttt{THEN} \quad X5 = Filtered\,Value \quad \texttt{ELSE} \quad X5 = f(X4) \tag{9.7}$$

- *X5_obs* is a pure clone of *X5*, i.e. *X5* is fully observed.

$$X5_{obs} = X5 \tag{9.8}$$

For the sake of completeness, we present the **Monitors** of *X5* (left) and *X5_obs* (right) in Figure 9.7.

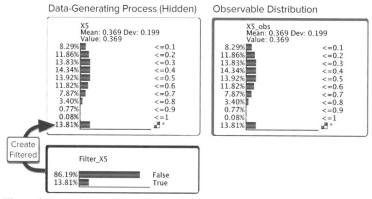

Figure 9.9

Missing Values Processing in BayesiaLab

BayesiaLab offers a wide range of missing values processing methods, which we are going to present in detail. However, we must emphasize that some of them, including listwise/casewise deletion and means imputation, are not recommended for default use. We still include these methods for two reasons: first, they are almost universally used in statistical analysis, and, secondly, under certain circumstances they can be safe to use. Regardless of their suitability for research, they highlight numerous challenges, which Bayesian networks and BayesiaLab can help overcome.

In the following, we will explore the advantages and disadvantages of the full range of techniques on the basis of a dataset that we are going to generate from our original, complete network (Figure 9.1). Given that we have specifically encoded all types of missingness mechanisms in this network, we can consider this dataset as a worst-case scenario, which is ideal for testing purposes.

Chapter 9

Generate Data

To begin this exercise, we have BayesiaLab produce the data that we will later use for testing. We can directly generate data according to the joint probability distribution encoded by the network (Figure 9.10): **Data > Generate Data**.

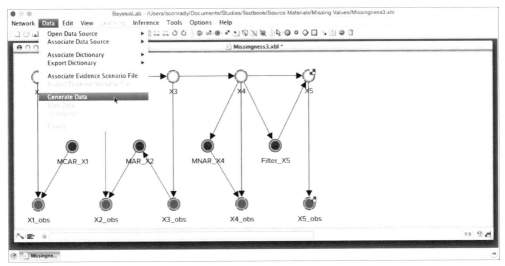

Figure 9.10

Next, we must specify whether to generate this data internally or externally (Figure 9.11). For now, we generate the data internally, which means that we associate data points with all nodes. This includes missing values and **Filtered Values** according to the original network (Figure 9.1). For the **Number of Examples** (i.e. cases or records), we set 10,000.

Figure 9.11

The database icon () signals that a dataset is now associated with the network. Additionally, we can see the number of cases in the database at the top of the **Monitor Panel** (Figure 9.12).

Now that this data exists inside BayesiaLab, we need to export it, so we can truly start "from scratch" with the test dataset. Also, in terms of realism, we only want to make the observable variables available, rather than all. We first select the nodes *X1_obs* through *X5_obs* and then select **Data > Save Data** from the main menu (Figure 9.12).

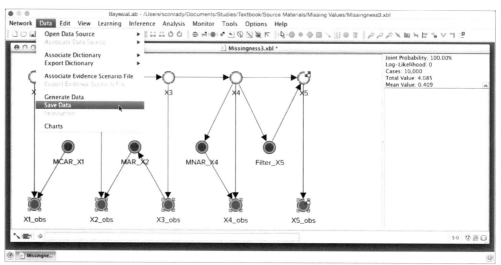

Figure 9.12

Next, we confirm that we only want to save the **Selected Nodes**, i.e. the observable variables (Figure 9.13).

Figure 9.13

Upon specifying a file name and saving the file, the export task is complete. A quick look at the CSV file confirms that the newly generated data contain both missing values and **Filtered Values**, as indicated with question marks and asterisks respectively (Figure 9.14).

X1_obs	X2_obs	X3_obs	X4_obs	X5_obs
0.3783	0.3918	0.4755	?	0.4325
0.1805	?	0.0013	0.0128	0.0285
?	0.3301	0.4012	0.3250	0.2648
0.4255	0.4228	0.3131	0.4745	0.5560
0.2417	?	0.2706	0.3522	0.3501
0.1871	0.2761	0.2408	0.2130	0.0755
0.2969	?	0.0282	0.1632	0.2486
?	?	0.3287	0.2853	0.2530
?	?	0.5532	0.6657	0.5450
0.3000	0.3828	0.3089	0.4449	0.3303
0.2696	0.2944	0.3908	0.3136	0.2844
0.5140	0.5324	0.4672	0.2563	0.0063
?	0.3462	0.3493	0.1822	0.2123
0.7138	0.8546	0.9913	0.9055	*
0.4547	0.4679	0.4244	0.5196	0.6340
?	?	0.3543	?	0.4009
?	0.7393	0.7520	0.8349	*
0.1101	?	0.1035	0.1218	0.2954
0.5611	0.3608	0.3052	0.3254	0.2020
0.1644	?	0.2596	0.3652	0.3655

Figure 9.14

Now that we have produced a dataset with all types of missingness, we discard our data-generating model and start "from scratch." We approach this dataset as if this were the first time we see it, without any assumptions and without any background knowledge. This provides us with a suitable test case for BayesiaLab's range of missing values processing methods.

Data Import Wizard

In a typical data analysis workflow in BayesiaLab, we first encounter **Missing Values Processing** in the **Data Import Wizard**. There, we need to choose from several options, which can be grouped into **Filter**, **Replace By**, and **Infer**. We will repeat the **Data Import Wizard** for each available **Missing Values Processing** option and then utilize each option in conjunction with the appropriate learning and estimation algorithms. This will yield one completed dataset for each **Missing Values Processing** approach in BayesiaLab. With these newly completed datasets, we can examine how well they match the original data. Ultimately, this gives us guidance for choosing the appropriate **Missing Values Processing** option as a function of what we know about the data-generating process and the missingness mechanism in particular.

Data Import

Our dataset consisting of 10,000 records was saved as a text file, so we start the import process via **Data > Open Data Source > Text File**. We show the first two steps of the **Data Import Wizard** only for reference as their options have already been discussed in previous chapters. Note the missing values in columns *X1_obs*, *X2_obs*, and *X4_obs* in the **Data** panel (Figure 9.15). Column *X5_obs* features **Filtered Values**, which are marked with an asterisk ("*").

Figure 9.15

The next step of the **Data Import Wizard** requires no further input, but we can review the statistics provided in the **Information Panel** (Figure 9.16): we have 5,547 missing values (=11.09% of all cells in the **Data** panel) and 1,364 **Filtered Values** (=2.73%).

Chapter 9

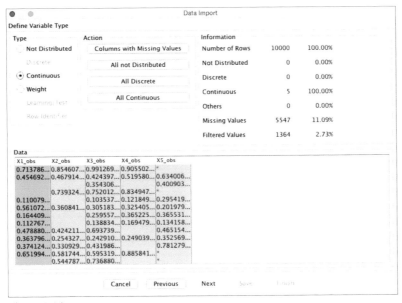

Figure 9.16

The screen of Figure 9.17 brings us to the core task of selecting the **Missing Values Processing** method. The default option[3] is pre-selected, but we will instead explore all options systematically from the top.

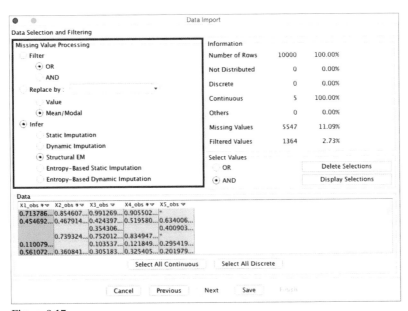

Figure 9.17

---

3 The default method can be specified under **Settings > Data > Import & Associate > Missing & Filtered Values**.

303

Filter (Listwise/Casewise Deletion)

BayesiaLab's **Filter**[4] method is generally known as "listwise deletion" or "casewise deletion" in the field of statistics. It is the first option listed, and it represents the simplest approach of dealing with missing values, and it is presumably the most commonly used one, too. This method simply deletes any record that contains a missing value in the specified variables.

Figure 9.18 shows **Filter** applied to *X1_obs* only. Given this selection, the **Number of Rows**, i.e. the number cases or records in the dataset, drops from the original 10,000 to 8,950. Note that **Filter** can be applied variable by variable. Thus, it is possible to apply **Filter** to a subset of variables only and use other methods for the remaining variables.

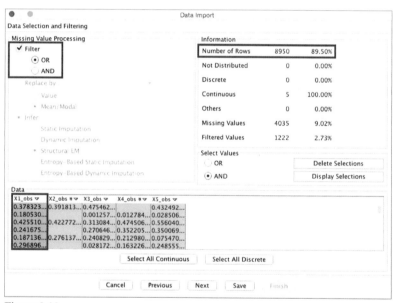

Figure 9.18

Before we can evaluate the effect of **Filter**, we need to complete the data import process. However, given the number of times we have already presented the entire import process, we omit a detailed presentation of these steps. Instead, we fast forward to review the **Monitors** of the processed variables in BayesiaLab (Figure 9.19).

---

[4] The **Filter** method is not to be confused with **Filtered Values** (see Filtered Values, p. 296.)

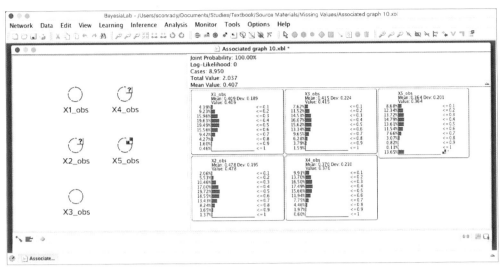

Figure 9.19

In the **Graph Panel**, the absence of the question mark icon on *X1_obs* signals that it no longer contains any missing values.

The **Monitors** now show the processed distributions. However, for a formal review of the processing effects, we need to compare the distributions of the newly processed variables with their unprocessed counterparts.

In Figure 9.20, we compare the original distributions (left column), followed by the distributions corresponding to the 10,000 generated samples (center column), and the distributions produced by the application of **Missing Values Processing** (right column). This is the format we will employ to evaluate all missing values processing methods.

Recalling the section on MCAR data, we know that applying **Filter** to an MCAR variable should not affect its distribution. Indeed, for *X1_obs* (top right) versus X1 (top left), the difference between the distributions is not significant, and it is only due to the sample size. Sampling an infinite size dataset would lead to the exact same distribution.

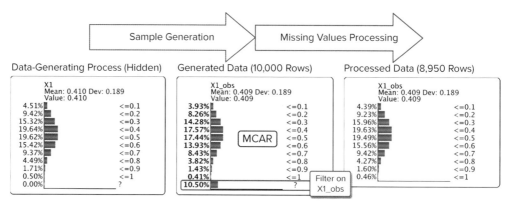

Figure 9.20

Now we turn to testing the application of **Filter** to *all* variables with missing values, i.e. *X1_obs*, *X2_obs*, and *X4_obs* (Figure 9.21).

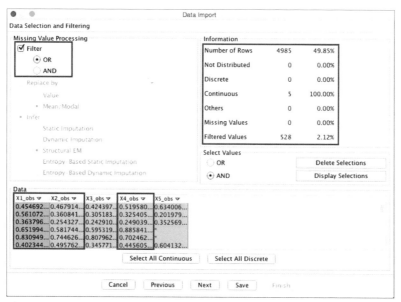

Figure 9.21

Even before evaluating the resulting distributions, we see in the **Information Panel** that over half of the rows of data are being deleted as a result of applying **Filter**. It is easy to see that in a dataset with more variables, this could quickly reduce the number of remaining records, potentially down to zero. In a dataset in which not a single record is completely observed, **Filter** is obviously not applicable at all.

Figure 9.22 presents the final distributions (right column), which are all substantially biased compared to the originals (left column). Whereas filtering alone on *X1_obs*, an MCAR variable, was at least "safe" for *X1_obs* by itself (Figure 9.20), filtering on *X1_obs*, *X2_obs*, and *X4_obs* adversely affects all variables, including *X1_obs* and even *X3_obs*, which does not contain any missing values.

As a result, we must strongly advise against using this method, within BayesiaLab or in any statistical analysis, unless there is certainty that all to-be-deleted incomplete observations correspond to missing values that have been generated completely at random (MCAR). Another exception would be if the to-be-deleted observations only represented a very small fraction of all observations. Unfortunately, these caveats are rarely observed, and the **Filter** method, i.e. listwise or casewise deletion, remains one of the most commonly used methods of dealing with missing values (Peugh and Enders, 2004).

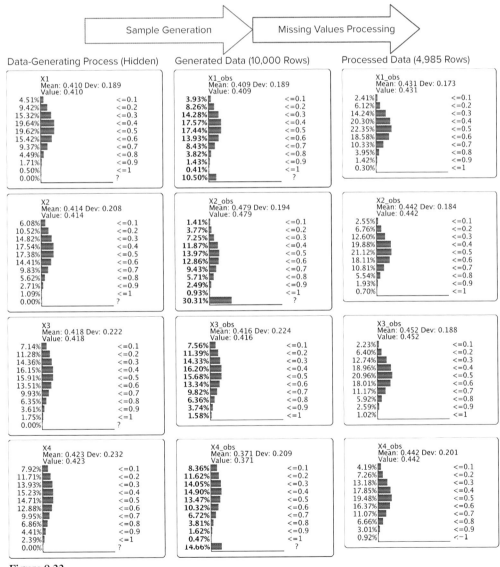

Figure 9.22

Replace By (Mean/Modal Imputation)

As opposed to deletion-type methods, such as **Filter**, we now consider the "opposite" approach, i.e. filling in the missing values with imputed values. Here, imputing means replacing the non-observed values with estimates, in order to facilitate the analysis of the whole dataset.

In BayesiaLab, this function is available via the **Replace By** option (Figure 9.23). We can specify to impute any arbitrary value, e.g. based on expert knowledge, or an automatically generated value. For a continuous variable, BayesiaLab offers a default replacement of the missing values with the mean value of the variable. For a discrete variable, the default is the modal value, i.e. the most frequently observed state of the variable. In our example, *X1_obs* has a mean value of 0.40878022. This is the value to be imputed for all missing values for *X1_obs*.

Note that **Replace By** can be applied variable by variable. Thus, it is possible to apply **Replace By** to a subset of variables only and use other methods for the remaining variables.

For the purposes of our example, we use **Replace By** for *X1_obs*, *X2_obs*, and *X4_obs*. As soon as this is specified, the number of the remaining missing values is updated in the **Information Panel**. By using the selected method, no missing values remain.

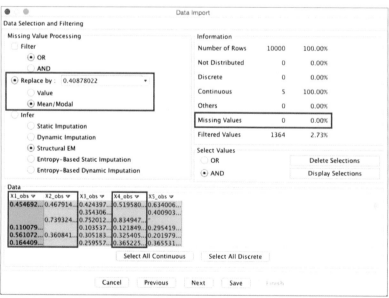

Figure 9.23

In the same way we studied the performance of **Filter**, we now review the results of the **Replace By** method (Figure 9.24). Whereas this imputation method is optimal at the individual/observation level (it is the rational decision for minimizing the prediction error), it is not optimal at the population/dataset level. The right column in Figure 9.24 shows that imputing all missing values with the same value has a strong impact on the shape of the distributions. Even though the mean values of the processed variables (right column) remain unchanged compared to observed values (center column), the standard deviation is much reduced.

Similar to our verdict on **Filter**, **Replace By** cannot be recommended either for general use. However, its application could be justified if expert knowledge were available for setting a specific replacement value or if the number of affected records were negligible compared to the overall size of the dataset.

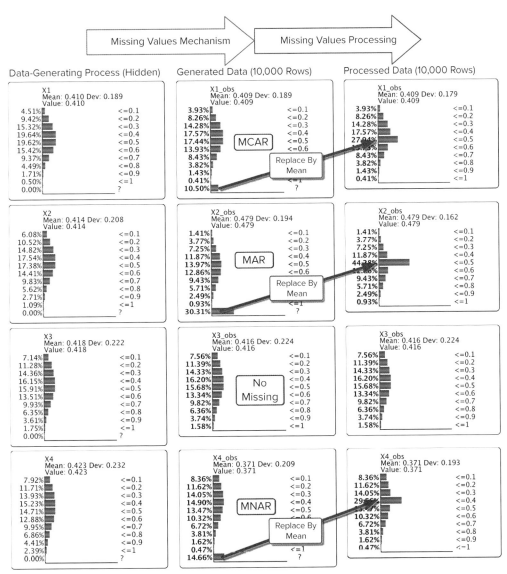

Figure 9.24

Infer: Static Imputation

Static Imputation resembles the **Replace By** method (i.e. mean/modal imputation) but differs in three important aspects:

1. While **Replace By** is deterministic, **Static Imputation** performs random draws from the marginal distributions of the observed data and saves these randomly-drawn values as "placeholder values."
2. The imputation is only performed internally, and BayesiaLab still "remembers" exactly which observations are missing.
3. Whereas **Replace By** can be applied to *individual* variables, any of the options under **Infer** apply to *all* variables with missing values, with the exception of those that have already been processed by **Filter** or **Replace By**.

The buttons under **Infer** are available whenever a variable with missing values (?) is selected in the **Data** panel (Figure 9.25).

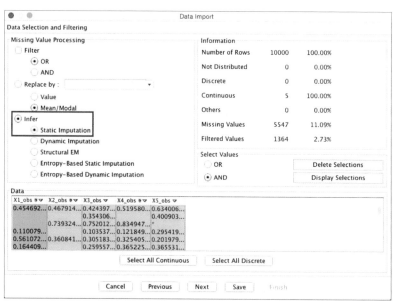

Figure 9.25

Although this probabilistic imputation method is not optimal at the observation/individual level (it is not the rational decision for minimizing the prediction error), it is optimal at the dataset/population level.

As illustrated in Figure 9.26, by drawing the imputed values from the current distribution keeps the distributions of variables pre- and post-processing the same. As a result, **Static Imputation** returns distributions that match the ones produced by **Filtering**, but without deleting any observations. As no records are discarded, **Static**

Chapter 9

Imputation does not introduce any additional biases. However, the distributions of $X2$ (MAR) and $X4$ (MNAR) remain strongly biased.

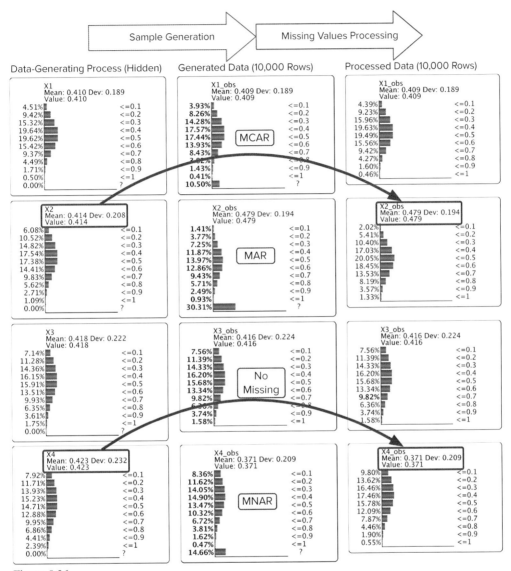

Figure 9.26

Infer: Dynamic Imputation

Dynamic Imputation is the first of a range of methods that take advantage of the structural learning algorithms available in BayesiaLab.

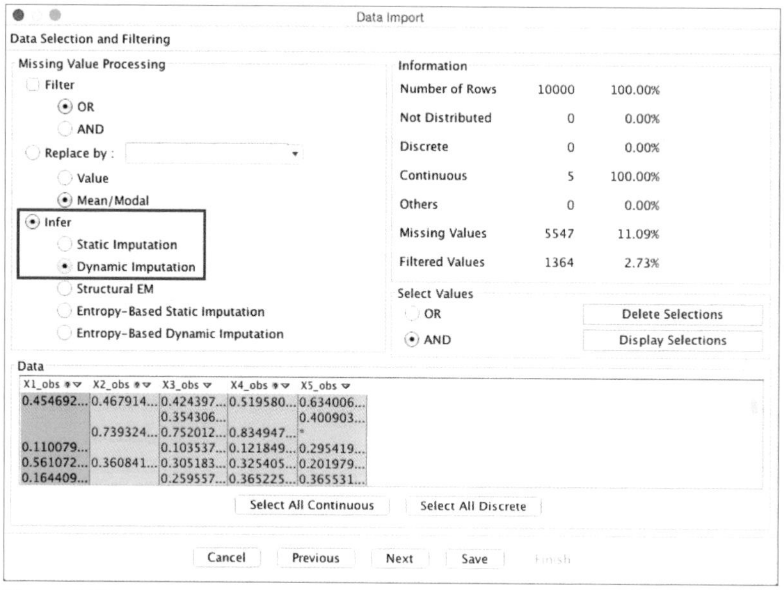

Figure 9.27

Like **Static Imputation**, **Dynamic Imputation** is probabilistic; imputed values are drawn from distributions. However, unlike **Static Imputation**, **Dynamic Imputation** does not only perform imputation once, but rather whenever the current model is modified, i.e. after each arc addition, deletion, and reversal during structural learning. This way, **Dynamic Imputation** always uses the latest network structure for updating the distributions from which the imputed values are drawn.

Upon completion of the data import, the resulting unconnected network initially has exactly the same distributions as the ones we would have obtained with **Static Imputation**. In both cases, imputation is only based on marginal distributions. With **Dynamic Imputation**, however, the imputation quality gradually improves during learning as the structure becomes more representative of the data-generating process. For example, a correct estimation of the MAR variables is possible once the network contains the relationships that explain the missingness mechanisms.

Dynamic Imputation might also improve the estimation of MNAR variables if the structural learning finds relationships with proxies of hidden variables that are part of the missingness mechanisms.

In Figure 9.28, the question marks (**?**) associated with *X1_obs*, *X2_obs*, and *X4_obs* confirm that the missingness is still present, even though the observations have been internally imputed.

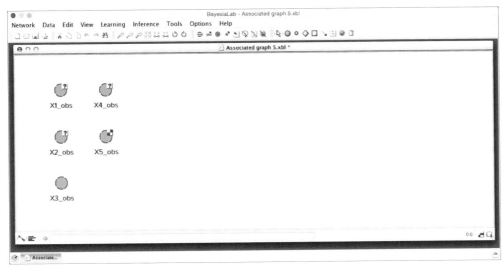

Figure 9.28

On the basis of this unconnected network, we can perform structural learning. We select **Learning > Unsupervised Structural Learning > Taboo** (Figure 9.29).

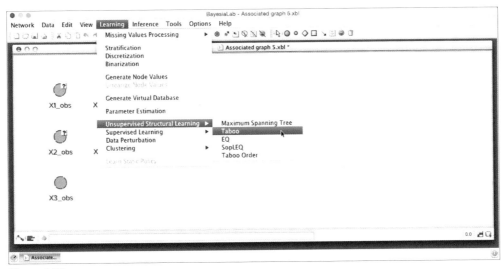

Figure 9.29

While the network (Figure 9.30) only takes a few moments to learn, we notice that it is somewhat slower compared to what we would have observed using a non-dynamic missing values processing method, e.g. **Filter**, **Replace By**, or **Static Imputation**. For our small example, the additional computation time requirement is immaterial. However, the computational cost increases with the number of variables in the network, the number of missing values, and, most importantly, with the complexity of the network. As a result, **Dynamic Imputation** can slow down the learning process significantly.

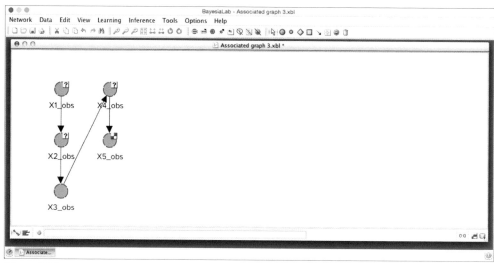

Figure 9.30

To make the network in Figure 9.30 visually consistent with the original order of the variables, we select all nodes and then select **Alignment > Horizontal Distribution** from the contextual menu (Figure 9.31).

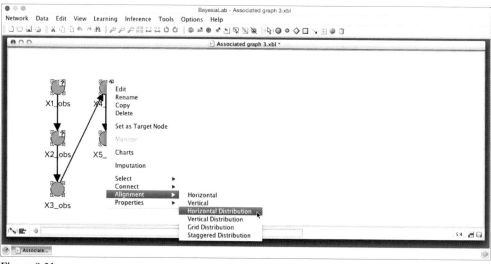

Figure 9.31

Figure 9.32 shows the horizontally aligned network.

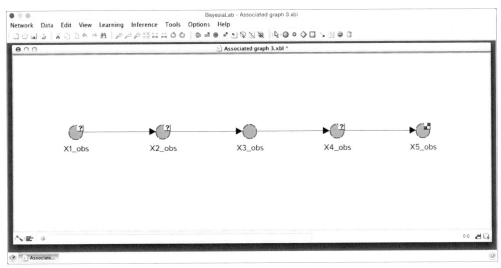

Figure 9.32

Figure 9.33 reports the performance of the **Dynamic Imputation**. The distributions show a substantial improvement compared to all the other methods we have discussed so far. As expected, *X2_obs* is now correctly estimated, and it even improves the distribution estimation of the difficult-to-estimate MNAR variable *X4_obs*. More specifically, there is now much less of an underestimation of the mean value.

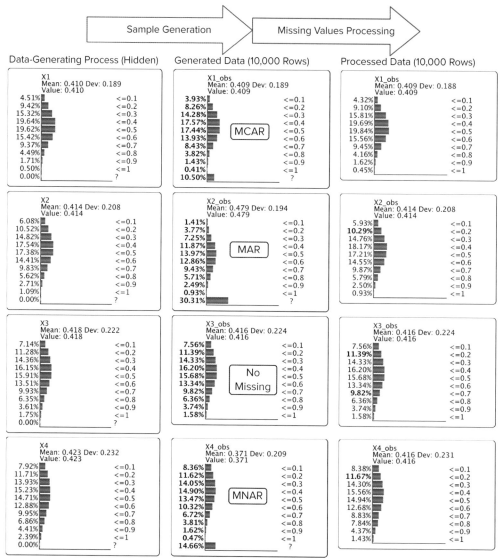

Figure 9.33

Infer: Structural EM

Structural Expectation Maximization (or **Structural EM** for short) is the next available option under **Infer** (Figure 9.34). This method is very similar to **Dynamic Imputation**, but instead of imputing values after each structural modification of the model, the set of observations is supplemented with one weighted observation per combination of the states of the jointly unobserved variables. Each weight equals the posterior joint probability of the corresponding state combination.

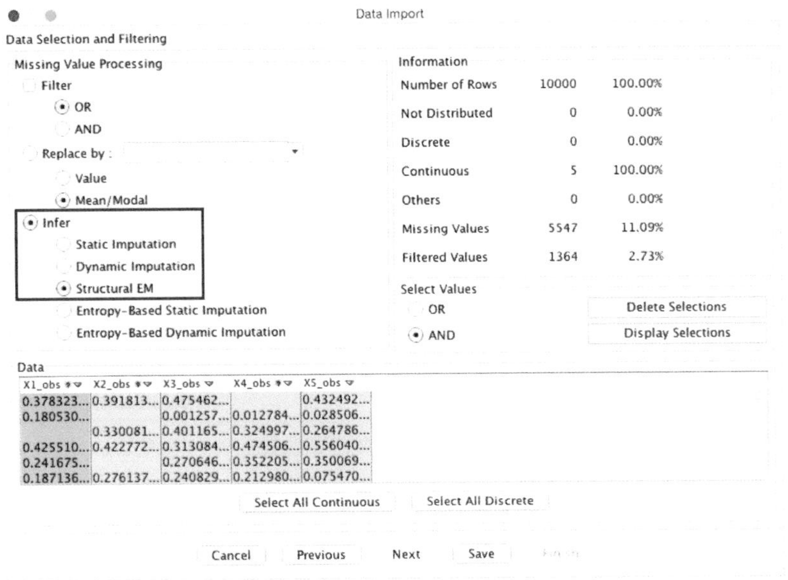

Chapter 9

Figure 9.34

Upon completion of the data import process, we perform structural learning again, analogously to what we did in the context of **Dynamic Imputation**. As it turns out, the discovered structure is equivalent to the one previously learned. Hence, we can immediately proceed to evaluate the performance (Figure 9.35).

The distributions produced by **Structural EM** are quite similar to those obtained with **Dynamic Imputation**. At least in theory, **Structural EM** should perform slightly better. However, the computational cost can be even higher than that of **Dynamic Imputation** because the computational cost of **Structural EM** also depends on the number of state combinations of the jointly unobserved variables.

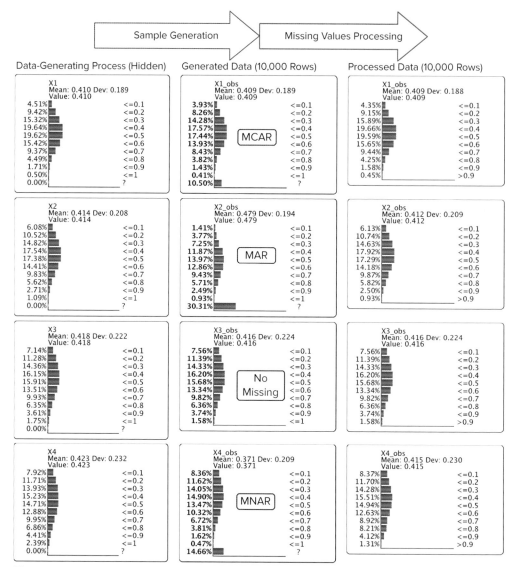

Figure 9.35

Entropy-Based Imputations

Under **Infer**, we have two additional options, namely **Entropy-Based Static Imputation** and **Entropy-Based Dynamic Imputation** (Figure 9.36). As their names imply, they are based on **Static Imputation** and **Dynamic Imputation**.

Chapter 9

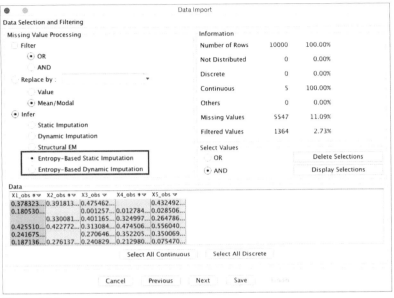

Figure 9.36

Whereas the standard (non-entropy-based) approaches randomly choose the sequence in which missing values are imputed within a row of data, the entropy-based methods select the order based on the conditional uncertainty associated with the unobserved variable. More specifically, missing values are imputed first for those variables that meet the following conditions:

1. Variables that have a fully-observed/imputed **Markov Blanket**
2. Variables that have the lowest conditional entropy, given the observations and imputed values

The advantages of the entropy-based methods are (a) the speed improvement over their corresponding standard methods and (b) their improved ability to handle datasets with large proportions of missing values.

Approximate Dynamic Imputation

As stated earlier, any substantial improvement in the performance of missing values processing comes at a high computational cost. Thus, we recommend an alternative workflow for networks with a large number of nodes and many missing values. The proposed approach combines the efficiency of **Static Imputation** with the imputation quality of **Dynamic Imputation**.

Static Imputation is efficient for learning because it does not impose any additional computational cost on the learning algorithm. With **Static Imputation**, missing values are imputed in-memory, which makes the imputed dataset equivalent to a fully observed dataset.

Even though, by default, **Static Imputation** runs only once at the time of data import, it can be triggered to run again at any time by selecting **Learning > Parameter Estimation**. Whenever **Parameter Estimation** is run, BayesiaLab computes the probability distributions on the basis of the current model. The missing values are then imputed by drawing from these distributions. If we now alternate structural learning and **Static Imputation** repeatedly, we can approximate the behavior of the **Dynamic Imputation** method. The speed advantage comes from the fact that values are now only imputed (on demand) at the completion of each full learning cycle as opposed to being imputed at every single step of the structural learning algorithm.

Approximate Dynamic Imputation in Practice

As a best-practice recommendation, we propose the following sequence of steps.

1. During data import, we choose **Static Imputation** (standard or entropy-based). This produces an initial imputation with the fully unconnected network, in which all the variables are independent.
2. We run the **Maximum Weight Spanning Tree** algorithm to learn a first network structure.
3. Upon completion, we prompt another **Static Imputation** by running **Parameter Estimation**. Given the tree structure of the network, pairwise

variable relationships provide the distributions used by the **Static Imputation** process.

4. Given the now-improved imputation quality, we start another structural learning algorithm, such as **EQ**, which may produce a more complex network.

5. The latest, more complex network then serves as the basis for yet another **Static Imputation**. We repeat steps 4 and 5 until we see the network converge towards a stable structure.

6. With a stable network structure in place, we change the imputation method from **Static Imputation** to **Structural EM** via **Learning > Missing Values Processing > Structural EM** (Figure 9.37).

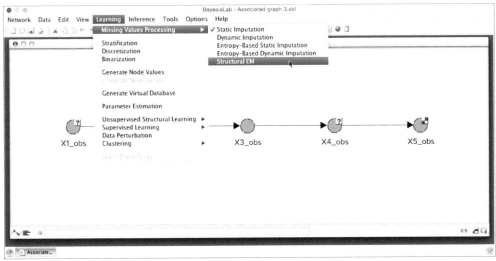

Figure 9.37

7. With the new imputation method set, running **Parameter Estimation** now prompts the **Structural EM** algorithm, which fine-tunes the imputation.

While this approximate **Dynamic Imputation** workflow requires more input and supervision by the researcher, for learning large networks, it can save a substantial amount of time compared to using the all-automatic **Dynamic Imputation** or **Structural EM**. Here, "substantial" can mean the difference between minutes and days of learning time.

Summary

Traditionally, the process of specifying an imputation model has been a scientific modeling effort on its own, and few non-statisticians dared to venture into this specialized field (van Buuren, 2007). With Bayesian networks and BayesiaLab, handling missing values properly now becomes feasible for researchers who might otherwise not attempt to deal with missing values beyond the ad hoc approaches. Responding to Burton and Altman's serious concern stated in the introduction of this chapter, we believe that the presented methods can help missing values processing become an integral part many more research projects in the future.

Chapter 9

10. Causal Identification & Estimation[1]

"Δημόκριτος έλεγε βούλεσθαι μάλλον μίαν ευρείν αιτιολογίαν ή την Περσών βασιλείαν εαυτού γενέσθαι." ("Democritus used to say that 'he prefers to discover a causality rather than become a king of Persia'.")—Democritus, according to a late testimony of Dionysius, Bishop of Alexandria, by Eusebius of Caesarea in Præparatio evangelica (Εὐαγγελικὴ προπαρασκευή)

The evolution of Bayesian network research is closely related to advances in the understanding of causality. In fact, both developments are intimately tied to the seminal works of Judea Pearl. Also, it is presumably fair to say that one of the "unique selling points" of Bayesian networks is their capability of performing causal inference. However, we do want to go beyond merely demonstrating the mechanics of causal inference. Rather, we rather want to establish under what conditions causal inference can be performed. More specifically, we want to see what assumptions are required to perform causal inference with non-experimental data.

To approach this topic, we need to break with the pattern established in the earlier chapters of this book. Instead of starting with a case study, we start off at a much higher level of abstraction. First, we discuss in theoretical terms as to what is required for performing causal identification, estimation, and inference. Once these fundamentals are established, we can proceed to discuss the methods, along with their limitations, including Directed Acyclic Graphs and Bayesian networks. These techniques can help us distinguish causation from association when working with non-experimental data.

---

[1] This chapter was prepared in collaboration with Felix Elwert on the basis of his course, Causal Inference with Graphical Models.

Motivation: Causality for Policy Assessment and Impact Analysis

In this chapter, we discuss causality mostly on the basis of a "toy problem," i.e. a simplified and exaggerated version of a real-world challenge. As such, the issues we raise about causality may appear somewhat contrived. Additionally, the constant practical use of causal inference in our daily lives may make our discussion seem somewhat artificial.

To highlight the importance of causal inference on a large scale, we want to consider how and under what conditions big decisions are typically made. Major government or business initiatives generally call for extensive studies to anticipate consequences of actions not yet taken. Such studies are often referred to as "policy analysis" or "impact assessment":

- "Impact assessment, simply defined, is the process of identifying the future consequences of a current or proposed action." (IAIA, 2009)
- "Policy assessment seeks to inform decision-makers by predicting and evaluating the potential impacts of policy options." (Adelle and Weiland, 2012)

What can be the source of such predictive powers? Policy analysis must discover a causal mechanism that links a proposed action/policy to a potential consequence/impact. Unfortunately, experiments are typically out of the question in this context. Rather, impact assessments—from non-experimental observations alone—must determine the existence and the size of a causal effect.

Given the sheer number of impact analyses performed, and their tremendous weight in decision making, one would like to believe that there has been a long-established scientific foundation with regard to (non-experimental) causal effect identification, estimation, and inference. Quite naturally, as decision makers quote statistics in support of policies, the field of statistics comes to mind as the discipline that studies such causal questions.

However, casual observers may be surprised to hear that causality has been anathema to statisticians for the longest time. "Considerations of causality should be treated as they always have been treated in statistics, preferably not at all..." (Speed, 1990).

The repercussions of this chasm between statistics and causality can be felt until today. Judea Pearl highlights this unfortunate state of affairs in the preface of his book Causality: "... I see no greater impediment to scientific progress than the prevailing practice of focusing all our mathematical resources on probabilistic and statistical inferences while leaving causal considerations to the mercy of intuition and good judgment." (Pearl, 1999)

Rubin (1974) and Holland (1986), who introduced the counterfactual (potential outcomes) approach to causal inference, can be credited with overcoming statisticians' traditional reluctance to engage causality. However, it will take many years for this fairly recent academic consensus to fully reach the world of practitioners, which is one of our key drivers for promoting Bayesian networks.

Sources of Causal Information

Causal Inference by Experiment

Randomized experiments have always been the gold standard for establishing causal effects. For instance, in the drug approval process, controlled experiments are mandatory. Without first having established and quantified the treatment effect, and any associated side effects, no new drug could win approval by the Federal Drug Administration or any other such organization.

Causal Inference from Observational Data and Theory

However, in many other domains, experiments are not feasible, be it for ethical, economical or practical reasons. For instance, it is clear that a government could not create two different tax regimes to evaluate their respective impact on economic growth. Neither would it be possible to experiment with two different levels of carbon emissions to measure the proposed warming effect.

"So, what does our existing data say?" would be an obvious question from policy makers, especially given today's high expectations concerning Big Data. Indeed, in lieu of experiments, we can attempt to find instances, in which the proposed policy already applies (by some assignment mechanism), and compare those to other instances, in which the policy does not apply.

However, as we will see in this chapter, performing causal inference on the basis of observational data requires an extensive range of assumptions, which can only come from theory, i.e. domain-specific knowledge. Despite all the wonderful advances in analytics in recent years, data alone, even Big Data, cannot prove the existence of causal effects.

Identification and Estimation Process

The process of determining the size of a causal effect from observational data can be divided into two steps:

Causal Identification

Identification analysis is about determining whether or not a causal effect can be established from the observed data. This requires a formal causal model, i.e. at least partial knowledge of how the data was generated. To justify any assumptions, domain knowledge is key. It is important to realize that the absence of causal assumptions cannot be compensated for by clever statistical techniques, or by providing more data. Needless to say, recognizing that a causal effect cannot be identified will bring any impact analysis to an abrupt halt.

Computing the Effect Size

If a causal effect is identified, the effect size estimation can be performed in the next step. Depending on the complexity of the model, this can bring a whole new set of challenges. Hence, there is a temptation to use familiar functional forms and estimators, e.g. linear models estimated by ordinary least squares (OLS). Beyond traditional approaches, we will exploit the properties of Bayesian networks in this context.

Theoretical Background

Today, we can openly discuss how to compute causal inference from observational data. For the better part of the 20th century, however, the prevailing opinion had been that speaking of causality without experiments is unscientific. Only towards the end of the century, this opposition had slowly eroded (Rubin 1974, Holland 1986), which subsequently led to numerous research efforts spanning philosophy, statistics, computer science, information theory, etc.

Chapter 10

Potential Outcomes Framework

Although there is no question about the common-sense meaning of "cause and effect", for a formal analysis, we require a precise mathematical definition. In the fields of social science and biostatistics, the potential outcomes framework[2] is a widely accepted formalism for studying causal effects. Rubin (1974) defines it as follows:

> "Intuitively, the causal effect of one treatment, $T=1$[3], over another, $T=0$, for a particular unit and an interval of time from t_1 to t_2 is the difference between what would have happened at time t_2 if the unit had been exposed to $T=1$ initiated at t_1 and what would have happened at t_2 if the unit had been exposed to $T=0$ initiated at t_1: 'If an hour ago I had taken two aspirins instead of just a glass of water, my headache would now be gone,' or because an hour ago I took two aspirins instead of just a glass of water, my headache is now gone.' Our definition of the causal effect of $T=1$ versus $T=0$ treatment will reflect this intuitive meaning."

- $Y_{i,1}$ Potential outcome of individual i given treatment $T=1$ (e.g. taking two Aspirins)
- $Y_{i,0}$ Potential outcome of individual i given treatment $T=0$ (e.g. drinking a glass of water)

The individual-level causal effect (ICE), is defined as the difference between the individual's two potential outcomes, i.e.

$$ICE = Y_{i,1} - Y_{i,0} \qquad (10.1)$$

Given that we cannot rule out differences between individuals (effect heterogeneity), we define the average causal effect (ACE), as the unweighted arithmetic mean of the individual-level causal effects:

$$ACE = E[Y_{i,1}] - E[Y_{i,0}] \qquad (10.2)$$

E[.] is the expected value operator, which computes the arithmetic mean.

---

2 The potential outcomes framework is also known as the counterfactual model, the Rubin model, or the Neyman-Rubin model.

3 In this quote from Rubin (1974), we altered the original variable name E to $T=1$ and C to $T=0$ in order to be consistent with the nomenclature in the remainder of this chapter. T is commonly used in the literature to denote the treatment condition.

Causal Identification

The challenge is that $Y_{i,1}$ (treatment) and $Y_{i,0}$ (non-treatment) can never be both observed for the same individual at the same time. We can only observe treatment or non-treatment, but not both.

So, where does this leave us? What we can produce easily is the "naive" estimator of association, S, between the "treated" and the "untreated"[4] sub-populations (for notational convenience we omit the index i):

$$S = E[Y_1 \mid T = 1] - E[Y_0 \mid T = 0] \tag{10.3}$$

Because the sub-populations in the treated and control groups contain different individuals, S is clearly not a measure of causation, in contrast to the ACE. This confirms the adage "association does not imply causation."

The question is, how can we move from what we can measure, i.e. the naive association, to the quantity of interest, i.e. causation? Determining whether we can extract causation from association, is known as identification analysis.

The safest approach of performing identification is conducting a randomized experiment. However, the premise of this chapter is that experiments are often not feasible for many research questions. Therefore, our only option is to see whether there were any conditions, under which the measure of association, S, equals the measure of causation, ACE. As a matter of fact, this would be the case if the sub-populations were comparable with respect to the factors that can influence the outcome.

Ignorability

Remarkably, the conditions under which we can identify causal effects from observational data are very similar to the conditions that justify causal inference in randomized experiments. A pure random selection of treated and untreated individuals does indeed remove any potential bias and allows estimating the effect of the treatment alone. This condition is known as "ignorability," which can be formally written as:

$$(Y_1, Y_0) \perp T \tag{10.4}$$

This means that the potential outcomes, Y_1 and Y_0 must jointly be independent ("$\perp$") of the treatment assignment, T. This condition of ignorability holds in an ideal experiment. Unfortunately, this condition is very rarely met in observational studies. How-

---

[4] In this chapter, we use "control", and "untreated" interchangeably.

ever, *conditional* ignorability may hold, which refers to ignorability within subgroups of the domain defined by the values of X.[5]

$$(Y_1, Y_0) \perp T \mid X \tag{10.5}$$

In words, conditional on variables X, Y_1 and Y_0 are jointly independent of T, the assignment mechanism. If conditional ignorability holds, we will be able to utilize the estimator, $S \mid X$, to recover the average causal effect, $ACE \mid X$.

$$\begin{aligned}
ACE \mid X &= E[Y_1 \mid X] - E[Y_0 \mid X] \\
&= E[Y_1 \mid T=1, X] - E[Y_0 \mid T=0, X] \\
&= E[Y \mid T=1, X] - E[Y \mid T=0, X] \\
&= S \mid X
\end{aligned} \tag{10.6}$$

How can we select the correct set of variables X among all variables in a system? How do we know that such variables X are observed, or even exist in a domain? The answer will presumably be unsatisfactory for many researchers and policy makers: it all depends on expert knowledge and assumptions.

Causal Assumptions

The MacMillan Dictionary defines "assumption" as "something that you consider likely to be true even though no one has told you directly or even though you have no proof." It is presumably fair to say that this carries a somewhat negative connotation. It implies that something is not known that perhaps should be known.

In some fields of science, assumptions can be perceived as a sign of weakness in reasoning. As a result, assumptions are often on the periphery of research projects, rather being at their core.

For causal identification with nonexperimental data, causal assumptions are crucial. More specifically, we must assert explicit causal assumptions about the process that generated the observed data (Manski 1999, Elwert 2013).

---

5   X can be a vector.

Example: Simpson's Paradox

We will use an example that appears trivial on the surface, but which has produced countless instances of false inference throughout the history of science. Due to its counterintuitive nature, this example has become widely known as Simpson's Paradox (Wall Street Journal, Dec. 2, 2009).

This is an important exercise as it illustrates how an incorrect interpretation of association can produce bias. The word "bias" may not necessarily strike fear into our hearts. In our common understanding, "bias" implies "inclination" and "tendency", and it is certainly not a particularly forceful expression. Hence, we may not be overly trouble by a warning about bias. However, Simpson's Paradox shows how bias can lead to catastrophically wrong estimates.

Does the Treatment Kill Patients?

A hypothetical disease equally affects men and women. An observational study finds that a treatment is linked to an increase in the recovery rate among all treated patients from 40 to 50% (Figure 10.1). Based on the study, this new treatment is widely recognized as beneficial and subsequently promoted as a new therapy.

	Patient Recovered	
Treatment	Yes	No
Yes	50%	50%
No	40%	60%

Figure 10.1

We can imagine a headline along the lines of "New Therapy Increases Recovery Rate by 10%." However, when examining patient records by gender, the recovery rate for male patients—upon treatment—decreases from 70% to 60%; for female patients, the recovery rate declines from 30% to 20% (Figure 10.2).

Gender	Treatment	Patient Recovered	
		Yes	No
Male	Yes	60%	40%
	No	70%	30%
Female	Yes	20%	80%
	No	30%	70%

Figure 10.2

Chapter 10

So, is this new treatment effective overall or not? This puzzle can be resolved by realizing that, in this observed population, there was an unequal application of the treatment to men and women, i.e. some type of self-selection occurred. More specifically, 75% of the male patients and only 25% of female patients received the treatment. Although the reason for this imbalance is irrelevant for inference, one could imagine that side effects of this treatment are much more severe for females, who thus seek alternatives therapies. As a result, there is a greater share of men among the treated patients. Given that men have a better a priori recovery prospect with this type of disease, the recovery rate of the all treated patients increases. So, what is the true causal effect of this treatment?

Synthetic Data

Our particular manifestation of Simpson's Paradox is not very far-fetched, but it is still hypothetical. Therefore, we must rely on synthetic data to make this problem domain tangible for our study efforts. We generate 1,000 observations by sampling from the joint probability distribution of the original DGP. Needless to say, for this dataset to be a suitable example for non-experimental observations, like we would find in them under real-world conditions, the true DGP is not known but merely an assumption. Our synthetic dataset consists of three variables with two discrete states each:[6]

▶ Generate Data in Chapter 9, p. 299.

- X1_Gender: Male (1)/Female (0)
- X2_Treatment: Yes (1)/No (0)
- X3_Outcome: Patient Recovered (1)/Patient Did Not Recover (0)

Figure 10.3 shows a preview of the first ten rows of the newly generated data.

X1_Gender	X2_Treatment	X3_Outcome
Female (0)	No (0)	Patient Did Not Recover (0)
Male (1)	Yes (1)	Patient Recovered (1)
Male (1)	Yes (1)	Patient Did Not Recover (0)
Female (0)	No (0)	Patient Did Not Recover (0)
Male (1)	Yes (1)	Patient Did Not Recover (0)
Female (0)	No (0)	Patient Recovered (1)
Female (0)	No (0)	Patient Recovered (1)
Male (1)	No (0)	Patient Recovered (1)
Female (0)	Yes (1)	Patient Did Not Recover (0)
Female (0)	Yes (1)	Patient Did Not Recover (0)

Figure 10.3

---

6   The dataset is available for download from the Bayesia website: www.bayesia.us/simpson

333

Methods for Identification and Estimation

For now, we set aside the dataset. We will return to it in each of the two workflows to be presented in this chapter, albeit at different points in the process. Both workflows have the objective of identifying and estimating causal effects from non-experimental data (Figure 10.4).[7]

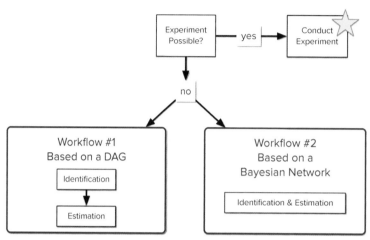

Figure 10.4

Workflow #1 uses a Directed Acyclic Graph for identification. In this approach, data will be introduced at the end of the process, only for estimation (Figure 10.5). Workflow #2 is based on a causal Bayesian network (CBN), which adds a JPD parametrization to a DAG for identification and estimation. In this context, we need to introduce data almost immediately (Figure 10.14).

Workflow #1: Identification and Estimation with a DAG

In this workflow (highlighted in Figure 10.5), all causal assumptions for identification are expressed explicitly in the form of a Directed Acyclic Graph (DAG) (Pearl 1995, 2009). These assumptions are not merely checkboxes to tick; rather, they represent our complete causal understanding of the data-generating process for the system we are studying. Where do we get such causal assumptions for a model? In this day and age, when Big Data dominates the headlines, we would like to say that advanced algorithms can generate causal assumptions from data. That is not the case, unfortunately. Structural, causal assumptions still require human expert knowledge, or, more gen-

---

[7] Throughout the workflows diagrams, a golden star marks experiments as the optimal approach, the "gold standard," for estimating causal effects.

erally, theory. In practice, this means that we need to build (or draw) a causal graph of our domain, which we can subsequently examine with regard to identification.

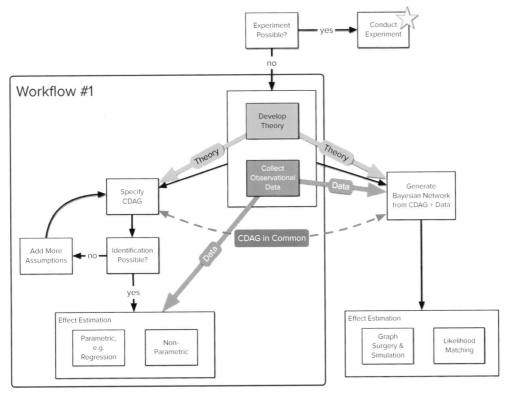

Figure 10.5

DAGs versus Bayesian Networks

As we learned in Chapter 2, Bayesian networks use DAGs for the qualitative representation of probabilistic dependencies. In the context of causal identification, however, the arcs in DAGs have a slightly different meaning. Here, arc direction explicitly states causality, as opposed to only representing a probabilistic dependency in a Bayesian network. To highlight this distinction, we introduce two new expressions: From now on, we refer to a non-causal DAG in a Bayesian network as a **Probabilistic DAG** (or **PDAG**). In contrast, we designate a DAG with a causal semantic as a **Causal DAG** (or **CDAG**). Given their importance for identification, we need to emphasize the theoretical properties of arcs in a **CDAG**:

▶ Chapter 2. Bayesian Network Theory, p. 21.

- A **Directed Arc** represents a potential causal effect. The arc direction indicates the assumed causal direction, i.e. "A → B" means "A causes B."
- A **Missing Arc** encodes the definitive absence of a direct causal effect, i.e. no arc between A and B means that there exists no direct causal relationship between A and B and vice versa. As such, a missing arc represents an assumption.

Structures Within a DAG

In a **PDAG** or **CDAG**, there are three basic configurations in which nodes can be connected. Graphs of any size and complexity can be broken down into these basic graph structures. While these basic structures show direct dependencies/causes explicitly, there are more statements contained in them, albeit implicitly. In fact, we can read all marginal and conditional associations that exist between the nodes.

Why are we even interested in associations? Isn't all this about understanding causal effects? It is essential to understand all associations in a system because, in non-experimental data, all we can do is observe associations, some which represent non-causal relationships. Our objective is to separate causal effects from non-causal associations.

Indirect Connection

The DAG in Figure 10.6 represents an indirect connection of A on B via C.

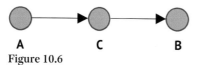

Figure 10.6

Implication for Causality

A causes B via node C.

Implication for Association

Marginally (or unconditionally), A and B are dependent. This means that without knowing the value of C, learning about A informs us about B and vice versa, i.e. the path between the nodes is unblocked and information can flow in both directions.

Conditionally on *C*, i.e. by setting **Hard Evidence**[8] on (or observing) *C*, *A* and *B* become independent. In other words, by "hard"-conditioning on *C*, we block the path from *A* to *B* and from *B* to *A*. Thus, *A* and *B* are rendered independent, given *C*: $A \not\perp B$ and $A \perp B | C$.

▸ Types of Evidence in Chapter 3, p. 42.

Common Parent

The second configuration has *C* as the common parent of *A* and *B* (Figure 10.7).

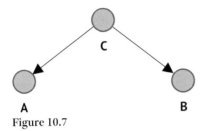

Figure 10.7

Implication for Causality

C it the common cause of both *A* and *B*

Implication for Association

In terms of association, this structure is absolutely equivalent to the Indirect Connection. Thus, *A* and *B* are marginally dependent, but conditionally independent given *C* (by setting **Hard Evidence** on *C*): $A \not\perp B$ and $A \perp B | C$

Common Child (Collider)

The final structure has a common child *C*, with *A* and *B* being its parents. This structure is called a "V-Structure." In this configuration, the common child *C* is also known as a "collider."

---

8   **Hard Evidence** means that there is no uncertainty with regard to the value of the observation or evidence. If uncertainty remains regarding the value of *C*, the path will not be entirely blocked and an association will remain between *A* and *B*.

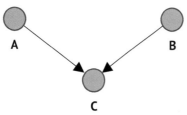

Figure 10.8

Implication for Causality

C is the common effect of A and B.

Implication for Association

Marginally, i.e. unconditionally, A and B are independent, i.e. the information flow between A and B is blocked. Conditionally on C—with any kind of evidence[9]—A and B become dependent. If we condition on the collider C, information can flow between A and B, i.e. conditioning on C opens the information flow between A and B: $A \perp B$ and $A \not\perp B | C$.

▶ Inter-Causal Reasoning in Chapter 4, p. 63.

For purposes of formal reasoning, there is a special significance to this type of connection. Conditioning on C facilitates inter-causal reasoning, often referred to as the ability to "explain away" the other cause, given that the common effect is observed.

Creating a CDAG Representing Simpson's Paradox

To model this problem domain, we create a simple CDAG with BayesiaLab, consisting of only three nodes, *X1_Gender*, *X2_Treatment*, and *X3_Outcome*.[10] The absence of further nodes means that we assume that there are no additional variables in the data-generating system, either observable or unobservable. This is a very strong assumption, which cannot be tested, unfortunately. To make such an assumption, we need to have a justification purely on theoretical grounds.

---

9 Even introducing a minor reduction of uncertainty of C, e.g. from no observation ("color unknown") to a very vague observation ("it could be anything, but it is probably not purple"), unblocks the information flow.

10 For now, we are only using the qualitative part of the network, i.e. we are not considering the probabilities.

Accepting this assumption for the time being, we wish to identify the causal effect of *X2_Treatment* on *X3_Outcome*. Is this possible by analyzing data from these three variables?

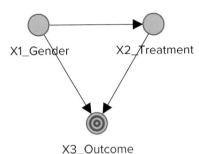

Figure 10.9

We need to ask, what does this CDAG specifically imply? We can find all three basic structures in this example:

- Indirect Effect: *X1_Gender* causes *X3_Outcome* via *X2_Treatment*
- Common Cause: *X1_Gender* causes *X2_Treatment* and *X3_Outcome*
- Common Effect: *X1_Gender* and *X2_Treatment* cause *X3_Outcome*

Graphical Identification Criteria

Earlier we said that we also need to understand all the associations in a system, so that we can distinguish between causation and association. This requirement will perhaps become clearer now as we introduce the concepts of causal and non-causal paths.

Causal and Non-Causal Paths

In a DAG, a path is a sequence of non-intersecting, adjacent arcs, regardless of their direction.
- A *causal* path can be any path from cause to effect, in which all arcs are directed away from the cause and pointed towards the effect.
- A *non-causal* path can be any path between cause and effect, in which at least one of the arcs is oriented from effect to cause.

Our example contains both types of paths:
1. Non-Causal Path: *X2_Treatment* ← *X1_Gender* → *X3_Outcome* (Figure 10.10).

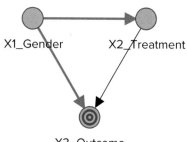
Figure 10.10

2. Causal Path: *X2_Treatment* → *X3_Outcome* (Figure 10.11).

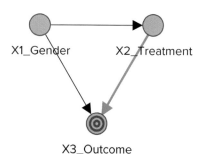
Figure 10.11

Adjustment Criterion and Identification

Among numerous available graphical criteria, the Adjustment Criterion (Shpitser et al., 2010) is perhaps the most intuitive one. Put simply, the Adjustment Criterion states that a causal effect is identified, if we can condition on (or adjust for) a set of nodes such that:

- All non-causal paths between treatment and effect are "blocked" (non-causal relationships prevented).
- All causal paths from treatment to effect remain "open" (causal relationships preserved).

This means that any association that we can measure after adjustment in our data must be causal, which is precisely what we wish to know. What does "adjust for" mean in practice? In this context, "adjusting for a variable" and "conditioning on a variable" are interchangeable. They can stand for any of the following operations, which all introduce information on a variable:

- Controlling
- Stratifying
- Setting evidence
- Observing
- "Mutilating"
- Matching

At this point, the adjustment technique is irrelevant. Rather, we just need to determine which variables, if any, need to be adjusted for in order to block the non-causal paths while keeping the causal paths open. Revisiting both paths in our DAG, we can now examine which ones are open or blocked.

- First, we look at the non-causal path in our DAG: *X2_Treatment* ← *X1_Gender* → *X3_Outcome* (Figure 10.10), i.e. *X1_Gender* is a common cause of *X2_Treatment* and *X3_Outcome*. This implies that there is an indirect association between *X2_Treatment* and *X3_Outcome*. Hence, there is an open non-causal path between *X2_Treatment* and *X3_Outcome*, which has to be blocked. To block this path, we simply need to adjust for *X1_Gender*.
- Next is the causal path in our DAG: *X2_Treatment* → *X3_Outcome* (Figure 10.11). It consists of a single arc from *X2_Treatment* to *X3_Outcome*, so it is open by default and cannot be blocked.

So, in this example, the Adjustment Criterion can be met by blocking the non-causal path *X2_Treatment* ← *X1_Gender* → *X3_Outcome* by adjusting for *X1_Gender*. Hence, the causal effect from *X2_Treatment* to *X3_Outcome* is identified.

Unobserved Variables

Thus far, we have assumed that there are no unobserved variables in our example. However, if we had reason to believe that there is another variable *U*, which appears to be relevant on theoretical grounds, but were not recorded in the dataset, identification could no longer be possible. Why? Let us assume *U* is a hidden common cause of *X2_Treatment* and *X3_Outcome*. By adding this unobserved variable *U*, a new non-causal path appears between *X2_Treatment* and *X3_Outcome* via *U* (Figure 10.12).

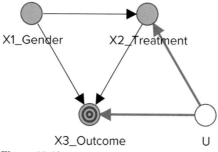

Figure 10.12

Given that U is unobserved, there is no way to adjust for it, and, therefore, this is an open non-causal path that cannot be blocked. Hence, the causal effect can no longer be estimated without bias. This highlights how easily identification can be "ruined."

Effect Estimation by Regression

Returning to the original version of the CDAG (Figure 10.9, i.e. without the unobserved variable), we are now ready to proceed to estimation. However, our CDAG, which helped us identify the causal effect, is only a qualitative representation of the data-generating process. To perform the actual effect estimation, we now need to quantify the relationships. This means that we have to specify a functional form and use our synthetic dataset for the effect estimation. A simple linear regression meets our requirements (we are assuming that there are no error terms):

▸ Synthetic Data, p. 333.

$$X3_{Outcome} = \beta_0 + \beta_1 X1_{Gender} + \beta_2 X2_{Treatment} \tag{10.7}$$

By estimating this regression, we condition on all the variables on the right-hand side of the equation (independent variables). With that, we have *X1_Gender* included as a covariate and, therefore, condition on it automatically. This is precisely what the adjustment criterion requires in our example. Based on the synthetic data, the OLS estimation yields the following coefficients:
$\beta_0 = 0.3, \quad \beta_1 = 0.4, \quad \beta_2 = -0.1$

Catastrophic Bias

We can now interpret the coefficient β_2 as the total causal effect of *X2_Treatment* on *X3_Outcome*. It turns out to be a *negative* effect. So, this causal analysis, which now removes bias by taking into account *X1_Gender*, yields the opposite effect of the one we would get by merely looking at association, i.e. −10% instead of +10% in recovery rate.

This illustrates that a bias in the estimation of an effect can be more than just a nuisance for the analyst. Bias can reverse the sign of the effect. In conditions similar to Simpson's Paradox, effect estimates can be substantially wrong and lead to policies with catastrophic consequences. In our example, the treatment under study would kill people, instead of healing them, as the naive study, based on association, first suggested.

Other Effects

Perhaps we are now tempted to interpret β_1 as the total causal effect of *X1_Gender* on *X3_Outcome*. This would not be correct. Instead, β_1 corresponds to the direct causal effect of *X1_Gender* on *X3_Outcome*. If we want to identify the total causal effect of *X1_Gender* on *X3_Outcome*, we will need to look once again at the paths in our DAG (Figure 10.13).

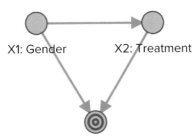

Figure 10.13

As it turns out, we have two causal paths from *X1_Gender* to *X3_Outcome*, and no non-causal path:

1. *X1_Gender* → X3_Outcome
2. *X1_Gender* → X2_Treatment → X3_Outcome

As a result, we must not adjust for *X2_Treatment*, because otherwise we would block the second causal path. A regression that included *X2_Treatment* would condition on *X2_Treatment* and thus block it. In order to obtain the total causal effect, a regression would have to be specified as in equation (10.8):

$$X3_{Outcome} = \beta_0 + \beta_1 X1_{Gender} \qquad (10.8)$$

Estimating the parameter yields $\beta_1 = 0.35$. Note that this illustrates that it is impossible to assign any causal meaning to regression coefficients without having an explicitly stated causal structure.

Workflow #2: Effect Estimation with Bayesian Networks

Conceptual Overview

In workflow #1, the causal effect estimation consisted of two separate elements: first, a CDAG that represented the qualitative part of the DGP, and, second, a classical regression that quantified the relationships and performed the effect estimation.

In workflow #2, we present a much more integrated approach by using a causal Bayesian network, which combines a CDAG with parameter estimates. This is the workflow we describe in this section (highlighted in Figure 10.14).

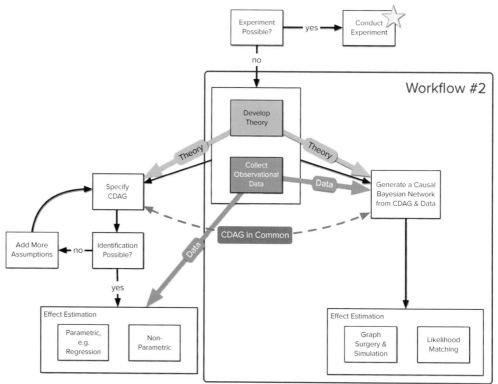

Figure 10.14

Creating a Causal Bayesian Network

We have already defined a CDAG in workflow #1, which we can reuse here towards building a causal Bayesian network. Figure 10.15 shows the familiar CDAG in Bayesia-Lab's **Graph Panel**. However, we still have to define the conditional probabilities for completing the specification of the causal Bayesian network. The yellow warning symbols (⚠) remind us that the parameters have not yet been defined.

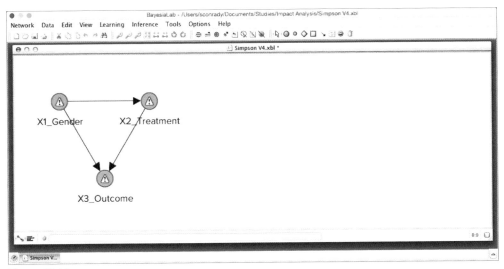

Figure 10.15

At this point, we could define the parameters based on our knowledge of all the probabilities in this domain. We took this approach in the knowledge modeling exercise of Chapter 5. Instead, we will utilize our synthetic dataset and BayesiaLab's **Parameter Estimation** to establish the quantitative part of the network.

▸ Synthetic Data, p. 333.

Associate Data

We have been using **Parameter Estimation** extensively in this book, either implicitly or explicitly, for instance in the context of structural learning and missing values estimation. So far, we have acquired the data needed for **Parameter Estimation** via the **Data Import Wizard**. Now will use the **Associate Data Wizard** for the same purpose. Whereas the **Data Import Wizard** generates *new* nodes from columns in a database, the **Associate Data Wizard** links columns of data with **existing** nodes. This way, we can "fill" our network with data and then perform **Parameter Estimation**.

▸ Parameter Estimation in Chapter 5, p. 99.

We start the **Associate Data Wizard** via the **Associate Data Source** function, which is available from the main menu under **Data > Associate > Data Source > Text File** (Figure 10.16).

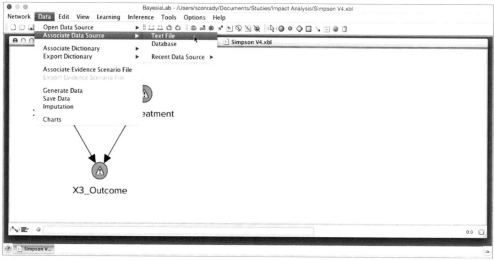

Figure 10.16

This prompts us to select the CSV file containing the our synthetic dataset. Upon selecting the file, BayesiaLab brings up the first screen of the **Associate Data Wizard** (Figure 10.17).

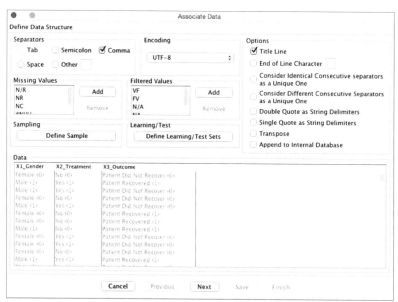

Figure 10.17

Given that the **Associate Data Wizard** mirrors the **Data Import Wizard** in most of its options, we omit to describe them here. We merely show the screens for reference as we click next to progress through the wizard (Figure 10.18 and Figure 10.19).

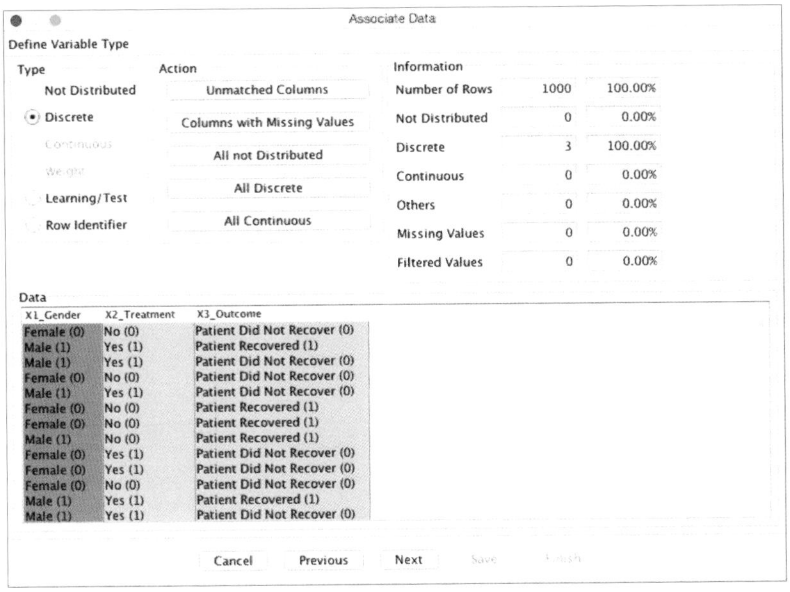

Figure 10.18

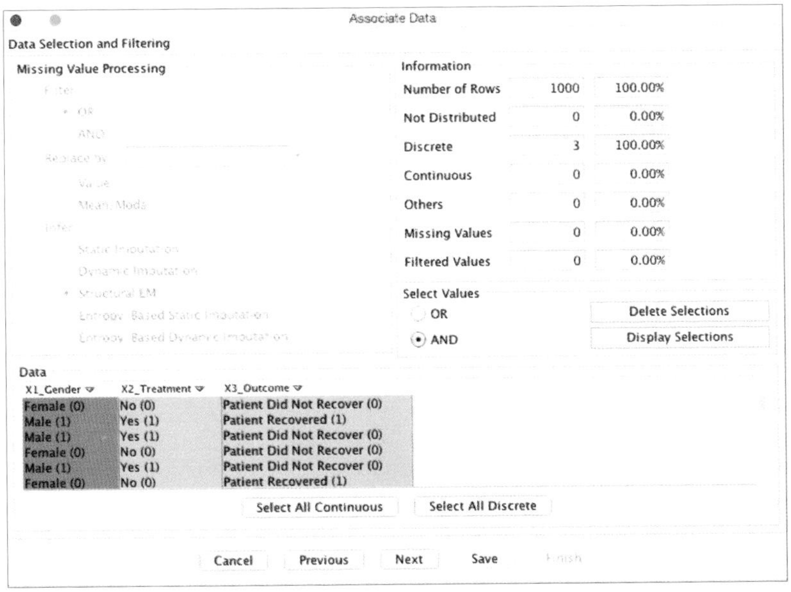

Figure 10.19

The last step shows how the columns in the dataset are linked to the nodes that already exist in the network (Figure 10.20). Conveniently, the column names in the dataset perfectly match the node names. Thus, BayesiaLab automatically associates the correct variables. If they did not match, we could manually link them in the final step (Figure 10.20).

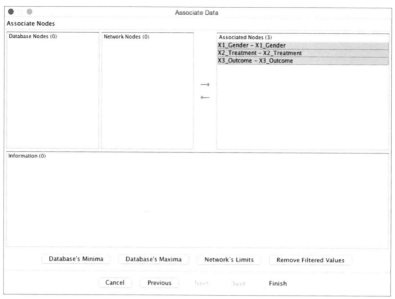

Figure 10.20

Clicking **Finish** completes the **Associate Data Wizard**. The database icon (⬚) indicates that our network now has a database associated with its structure. We can now use this database to estimate the parameters: **Learning > Parameter Estimation** (Figure 10.21).

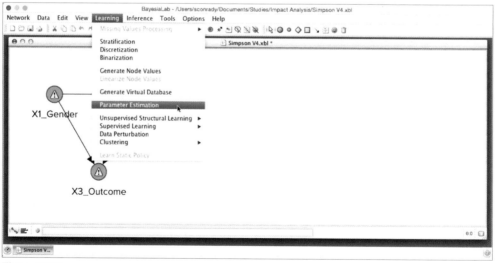

Figure 10.21

Once the parameters estimated, there are no longer any warning symbols (⚠) tagged onto the nodes (Figure 10.22).

Chapter 10

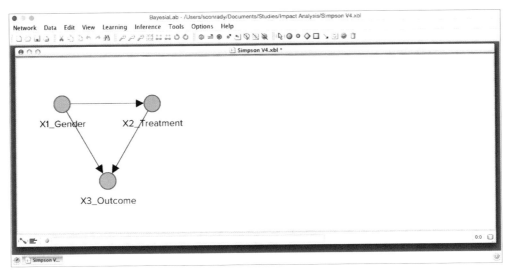

Figure 10.22

We now have a fully specified and estimated Bayesian network. By opening, for instance, the **Node Editor** of *X3: Outcome*, we see that the CPT is indeed filled with probabilities.

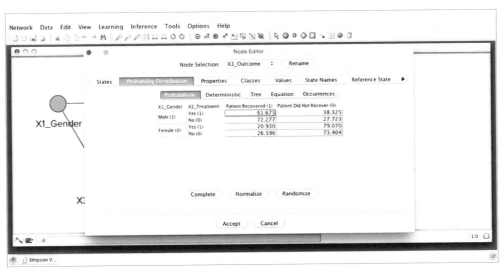

Figure 10.23

Path Analysis

Given that we now have a complete Bayesian network, BayesiaLab can help us understand the implications of the structure of this network. For instance, we can verify the paths in the network. Once we define a **Target Node**, we can examine the possible paths in this network. We select *X2_Treatment*, and then select **Analysis > Visual > Influence Paths to Target** (Figure 10.24).

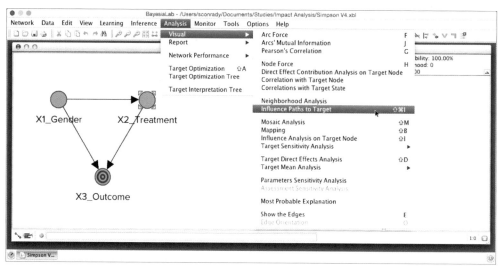

Figure 10.24

BayesiaLab then provides a pop-up window with the **Influence Paths** report. Selecting any of the listed paths shows the corresponding arcs in the **Graph Panel.** Causal paths are shown in blue; non-causal paths are pink (Figure 10.25).

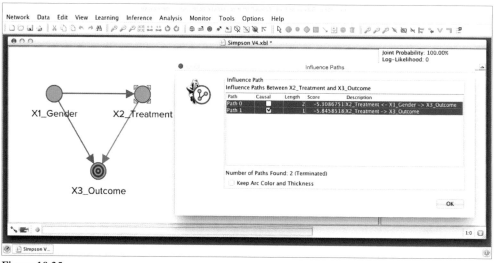

Figure 10.25

It is easy to see that this automated path analysis could be particularly helpful with more complex networks. In any case, the result confirms our previous, manual analysis, which means that we need to adjust for *X1_Gender* to block the non-causal path between *X2_Treatment* and *X3_Outcome* (pink path).

Chapter 10

Inference

To prepare this network for inference, we bring up the **Monitors** of all three nodes (Figure 10.26).

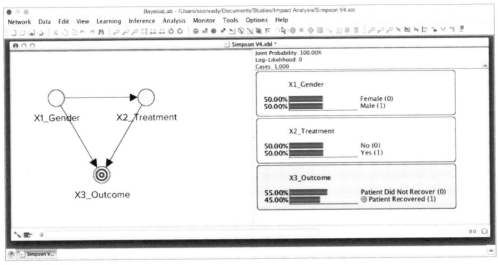

Figure 10.26

For instance, Figure 10.27 shows the prior distributions (left) and the posterior distributions (right), given the observation *X2_Treatment="Yes (1)"*.

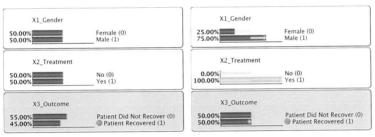

Figure 10.27

As one would expect, the target variable, *X3_Outcome*, changes upon setting this evidence. However, *X1_Gender*, changes as well, even though we know that this treatment could not possibly change the gender of a patient. In fact, what we observe here is a manifestation of the non-causal path: *X2_Treatment ← X1_Gender → X3_Outcome*. This is the very path we need to block, as per our earlier studies of the DAG, in order to estimate the causal effect, *X2_Treatment → X3_Outcome*.

So, how do we block a path in a Bayesian network? We do have a two options in this regard, and both are conveniently implemented in BayesiaLab.

Pearl's Graph Surgery

The concept of "graph surgery" is much more fundamental than our technical objective of blocking a path, as stipulated by the Adjustment Criterion. Graph surgery is based on the idea that a causal network represents a multitude of autonomous relationships between parent and child nodes in a system. Each node is only "listening" to its parent nodes, i.e. the child node's values are only a function of the value of its parents, not of any other nodes in the system. Also, these relationships remain invariant regardless of any values that other nodes in the network take on.

Should a node in this system be subjected to an outside intervention, the natural relationship between this node and its parents would be severed. This node no longer naturally "obeys" inputs from its parent nodes; rather an external force fixes the node to a new value, regardless of what the values of the parent nodes would normally dictate. Despite this particular disruption, the other parts of the network remain unaffected in their structure.

How does this help us estimate the causal effect? The idea is to consider the causal effect estimation as simulated interventions in the given system. Removing the arcs going into *X2_Treatment* implies that all the non-causal paths between *X2_Treatment* and the effect, *X3_Outcome*, no longer exist, without blocking the causal path (i.e. the same conditions apply as with the Adjustment Criterion).

Previously we computed the association in a system and interpreted it causally. Now have a causal network as a computational device, i.e. the Bayesian network, and can simulate what happens upon application of the cause. Applying the cause is the same as an intervention on a node in the network.

In our example, we wish to determine the effect of *X2_Treatment*, our cause, on *X3_Outcome*, the presumed effect. In its natural state, *X2_Treatment*, is a function of its sole parent *X1_Gender*. To simulate the cause, we must intervene on *X2_Treatment* and set it to specific values, i.e. "*Yes (1)*" or "*No (0)*", regardless of what *X1_Gender* would have induced. This is equivalent to randomly splitting the patient population into two sub-populations of equal size and forcing the first group to receive treatment and withholding the treatment from the second group. Such a random selection removes the association between *X2_Treatment* and *X1_Gender*. This severs the inbound arc from *X1_Gender* into *X2_Treatment*, as if it were "surgically" removed. However, all other properties remain unaffected, i.e. the distribution of *X1_Gender*, the arc between *X1_Gender* and *X3_Outcome*, and the arc between

X2_Treatment and *X3_Outcome*. This means, after performing the graph surgery, setting *X2_Treatment* to any value is an intervention, and any effects must be causal.

While we could perform graph surgery manually on the given network, this function is automated in BayesiaLab. After right-clicking the **Monitor** of the node *X2_Treatment*, we select **Intervention** from the **Contextual Menu** (Figure 10.28).

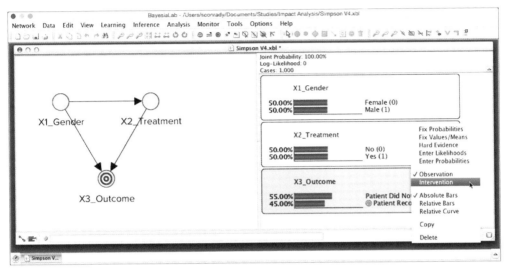

Figure 10.28

The activation of the **Intervention Mode** for this node is now highlighted by the blue background of the **Monitor** of *X2_Treatment* (Figure 10.29).

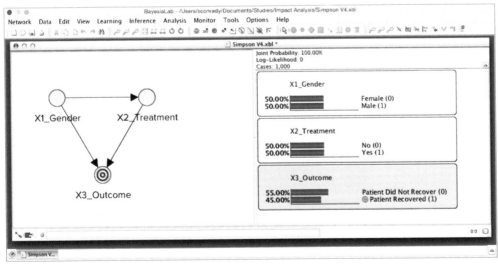

Figure 10.29

Setting evidence on *X2_Treatment* is now an **Intervention** and no longer an observation (Figure 10.30).

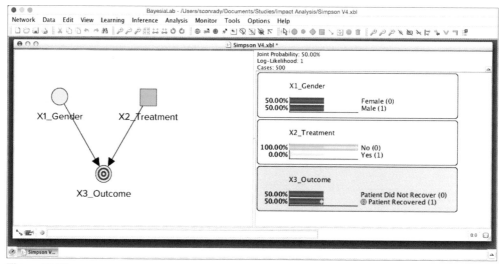

Figure 10.30

By setting the **Intervention**, BayesiaLab removes the inbound arc into *X2_Treatment* to visualize the graph mutilation (Figure 10.30). Additionally, the node symbol changes to a square (☐), which denotes a **Decision Node** in BayesiaLab. Furthermore, the distribution of *X1_Gender* remains unchanged. We first set *X2_Treatment*="No (0)", then we set *X2_Treatment*="Yes (1)", as shown in the **Monitors** (Figure 10.31).

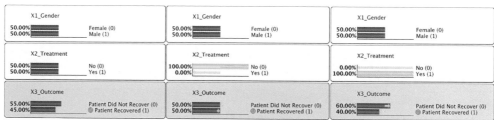

Figure 10.31

More formally, we can express these interventions with the do-operator.

$P(X3\_Outcome="Patient\ Recovered\ (1)"\ |\ do(X2\_Treatment="No(0)"))=0.5$

$P(X3\_Outcome="Patient\ Recovered\ (1)"\ |\ do(X2\_Treatment="Yes(1)"))=0.4$

As a result, the causal effect is −0.1.

As an alternative to manually setting the values of the intervention, we can employ BayesiaLab's **Total Effects on Target** function (Figure 10.32).

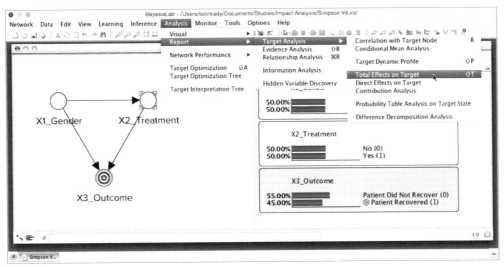

Figure 10.32

Given that we have set *X2_Treatment* to **Intervention Mode**, **Total Effects on Target** computes the total causal effect. Please note the arrow symbol (→) in the results table (Figure 10.33). This indicates that the **Intervention Mode** was active on *X2_Treatment*.

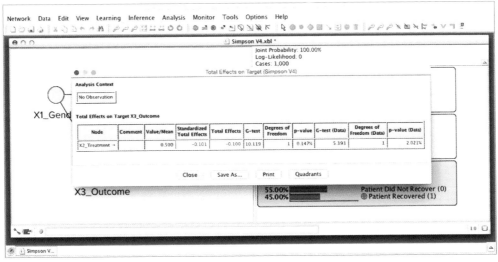

Figure 10.33

Introduction to Matching

Earlier in this chapter, adjustment was achieved by including the relevant variables in a regression. Instead, we now perform adjustment by matching. In statistics, match-

▸ Adjustment Criterion and Identification, p. 340.

ing refers to the technique of making distributions of the sub-populations we are comparing, including multivariate distributions, as similar as possible to each other. Applying matching to a variable qualifies as adjustment, and, as such, we can use it with the objective of keeping causal paths open and blocking non-causal paths. In our example, matching is fairly simple as we only need to match a single binary variable, i.e. *X1_Gender*. That will meet our requirement for adjustment and block the only non-causal path in our model.

Intuition for Matching

As the DAG-related terminology, e.g., "blocking paths", may not be universally understood by a non-technical audience, we can offer a more intuitive interpretation of matching, which our example can illustrate very well. We have seen that, because of the self-selection phenomenon we described in this population, by setting an observation on *X2_Treatment*, the distribution of *X1_Gender* changes. What does this mean? This means that given we observe those who are actually treated, i.e. *X2_Treatment="Yes (1)"*, they turn out to be 75% male. Setting the observation to "not treated", i.e. *X2_Treatment="No (0)"*, we only have a 25% share of male patients (Figure 10.34).

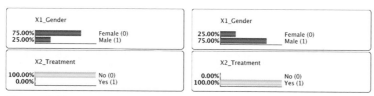

Figure 10.34

Given this difference in gender composition, comparing the outcome between the treated and the non-treated is certainly not an apples-to-apples comparison as we know from our model that *X1_Gender* also has a causal effect on *X3: Outcome*. Without controlling *X1_Gender*, the effect of *X2_Treatment* is confounded by *X1_Gender*.

So, how about searching for a subset of patients, in both treated and non-treated groups, which had an identical gender mix as illustrated in Figure 10.35 in order to neutralize the gender effect?

Chapter 10

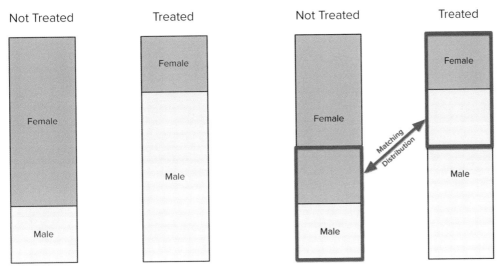

Figure 10.35

In statistical matching, this process typically involves the selection of units in such a way that comparable groups are created, as shown in Figure 10.36. In practice, this is typically a lot more challenging as the observed units have more than just a single binary attribute.

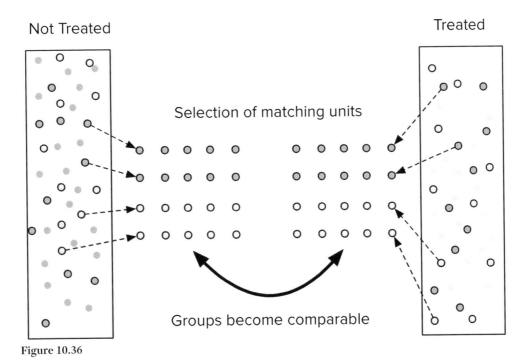

Figure 10.36

This approach can be extended to higher dimensions, meaning that the observed units need to be matched on a range of attributes, often including both continuous

357

and discrete variables. In that case, exact matching is rarely feasible, and some similarity measures must be utilized to define a "match."

Jouffe's Likelihood Matching

With **Likelihood Matching**, as it is implemented in BayesiaLab, however, we do not directly match the underlying observations. Rather we match the distributions of the relevant nodes on the basis of the joint probability distribution represented by the Bayesian network.

In our example, we need to ensure that the gender compositions of untreated (left) and treated groups (right) are the same, i.e. a 50/50 gender mix. This theoretically ideal condition is shown in the **Monitors** in Figure 10.37.

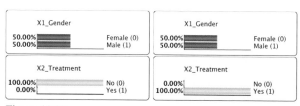

Figure 10.37

However, in Figure 10.38, the actual distributions reveal the inequality of gender distributions for the untreated (left) and the treated (right).

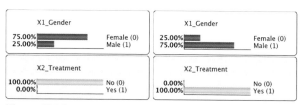

Figure 10.38

How can we overcome this? Consider that prior distributions exist for the to-be-matched variable *X1_Gender*, which, upon setting evidence on *X2_Treatment*, meet the desired, matching posterior distributions. In statistical matching, we would pick units that match upon treatment. In **Likelihood Matching**, however, we pick prior distributions that, upon treatment, have matching posterior distributions. In practice, for **Likelihood Matching**, "picking prior distributions" translates into setting **Probabilistic Evidence**.

Trying this out with actual distributions perhaps makes it easier to understand. We can set **Probabilistic Evidence** on the node *X1_Gender* by right-clicking on the **Monitor** and selecting **Enter Probabilities** from the contextual menu (Figure 10.39).

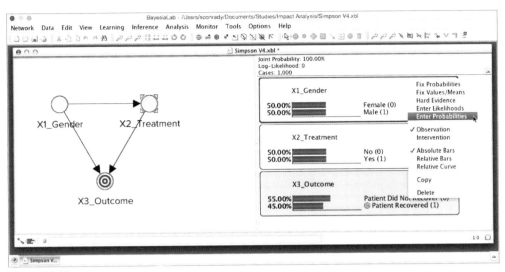

Figure 10.39

Now we can enter any arbitrary distribution for this node. For reasons that will become clear later, we set the distribution to 75% for *Male (1)*, which implies 25% for *Female (0)*. Given the new evidence, we also see a new distribution for *X2_Treatment* (Figure 10.40).

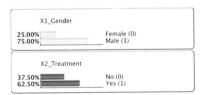

Figure 10.40

What happens now if we set treatment to *X2_Treatment="No (0)"*? As it turns out, *X1_Gender* assumes the very distribution that we desired for the treated group. Similarly, we can set probabilistic evidence on *X1_Gender* in such a way that *X2_Treatment= "Yes (1)"*, will also produce the 50/50 distribution. Hence, we have matching distributions for the untreated and the treated groups (Figure 10.41).

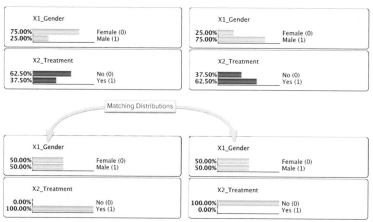

Figure 10.41

The obvious follow-up question would be how the appropriate probabilistic evidence can be found? In the example shown in Figure 10.41, we simply picked one without explanation, and it happened to produce the desired result. We will not answer this question, as the algorithm that produces the sets of probabilistic evidence is proprietary. However, for practitioners, this should be of little concern. **Likelihood Matching** is a fully-automated function in BayesiaLab, which performs the search in the background, without requiring any input from the analyst.

Direct Effects Analysis

So, what does this approach look like when applied to our example? We select *X2_Treatment* and the select **Analysis > Report > Target Analysis > Direct Effects on Target** (Figure 10.42).

Chapter 10

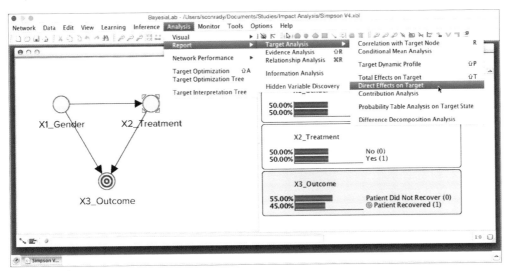

Figure 10.42

We immediately obtain a report that shows a **Direct Effect** of −0.1 (Figure 10.43).

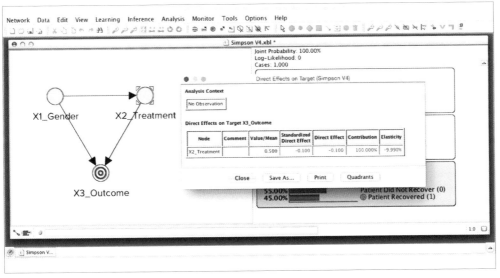

Figure 10.43

In BayesiaLab's terminology, a **Direct Effect** is the estimate of the effect between a node and a **Target Node**, by controlling for all variables that have not been defined as **Non_Confounder**.[11] In the current example, we only examined a single causal effect, but the **Direct Effects Analysis** can be applied to multiple causes in a single step.

---

11 This is intentionally aligned with the terminology used in the social sciences (Elwert, 2013).

Nonlinear Causal Effects

Due to the binary nature of all variables, our example was inherently linear. Hence, computing a single coefficient for the **Direct Effect** is adequate to describe the causal effect. However, the nonparametric nature of Bayesian networks offers another way of examining causal effects. Instead of estimating merely one coefficient to describe a causal effect, BayesiaLab can compute a causal "response curve." For reference, we now show how to perform a **Target Mean Analysis**. Instead of computing a single coefficient, this function computes the effect of interventions across a range of values. This function is available under **Analysis > Visual > Target Mean Analysis > Direct Effects** (Figure 10.44).

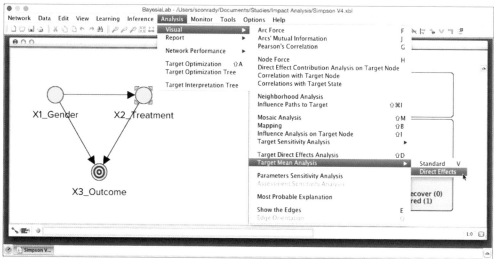

Figure 10.44

This brings up a pop-up window prompting us to select the format of the output. Selecting **Mean** for **Target**, and **Mean** for **Variables** is appropriate for this example (Figure 10.45).

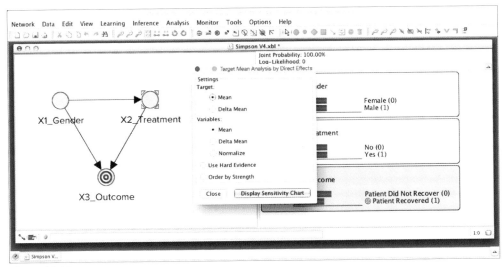

Figure 10.45

We confirm the selection by clicking **Display Sensitivity Chart**. Given the many iterations of this example throughout this tutorial, the resulting plot is not surprising. It appears to be a linear curve with the slope equivalent to the previously estimated causal effect (Figure 10.46).

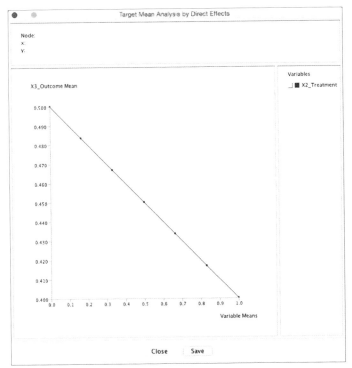

Figure 10.46

Soft Intervention

However, it is important to point out that it just *looks* like a linear curve. Loosely speaking, from BayesiaLab's perspective, the curve merely represents a series of points connected by lines. Each point was computed by setting **Interventions** at intermediate points between *X2_Treatment="No (0)"* and *X2_Treatment="Yes (1)"*. How should this be interpreted, given that *X2_Treatment* is a binary variable? The answer is that this can be considered as computing the causal effect of soft interventions.

▶ Inference with Probabilistic and Numerical Evidence in Chapter 7, p. 188.

In the context of policy analysis, this is perhaps highly relevant. One can certainly argue that many policies, when implemented, do rarely apply to all units. For instance, a nationwide vaccination program might only expect to reach 80% of the population. Hence, the treatment variable should presumably reflect that fact. Another example would be the implementation of a new speed limit. Once again, not all drivers will drive precisely at the speed limit. Rather, there is presumably a broad distribution of speeds, presumably centered roughly around the newly stipulated speed limit. So, simulating the real-world effect of an intervention requires us to compute it probabilistically, as shown here.

Conclusion

This chapter highlights how much effort is required to derive causal effect estimates from observational data. Simpson's Paradox illustrates how much can go wrong even in the simplest of circumstances. Given such potentially serious consequences, it is a must for policy analysts to formally examine all aspects of causality. To paraphrase Judea Pearl, we must not leave causal considerations to the mercy of intuition and good judgment.

It is fortunate that causality has emerged from its pariah status in recent decades, which has allowed tremendous progress in theoretical research and practical tools. "...practical problems relying on casual information that long were regarded as either metaphysical or unmanageable can now be solved using elementary mathematics" (Pearl, 1999).

Bayesian networks, and the BayesiaLab software platform are the direct result of this research progress. It is now upon the community of practitioners to embrace this progress to develop better policies, for the benefit of all of us.

Chapter 10

Bibliography

Adelle, Camilla, and Sabine Weiland. "Policy Assessment: The State of the Art." Impact Assessment and Project Appraisal 30, no. 1 (March 1, 2012): 25–33. doi:10.1080/14615517.2012.663256.

Bouhamed, Heni, Afif Masmoudi, Thierry Lecroq, and Ahmed Rebaï. "Structure Space of Bayesian Networks Is Dramatically Reduced by Subdividing It in Sub-Networks." Journal of Computational and Applied Mathematics 287 (October 2015): 48–62. doi:10.1016/j.cam.2015.02.055.

Breiman, Leo. "Statistical Modeling: The Two Cultures (with Comments and a Rejoinder by the Author)." Statistical Science 16, no. 3 (2001): 199–231.

Burton, A., and D. G. Altman. "Missing Covariate Data within Cancer Prognostic Studies: A Review of Current Reporting and Proposed Guidelines." British Journal of Cancer 91, no. 1 (July 5, 2004): 4–8. doi:10.1038/sj.bjc.6601907.

Conrady, Stefan, and Lionel Jouffe. "Artificial Intelligence for Research, Analytics, and Reasoning." Northwestern University, Evanston, Illinois, May 12, 2015. http://www.bayesia.us/bayesialab-101-northwestern-university.

Darwiche, Adnan. Modeling and Reasoning with Bayesian Networks. 1st ed. Cambridge University Press, 2009.

De Cock, Dean. "Ames, Iowa: Alternative to the Boston Housing Data as an End of Semester Regression Project." Journal of Statistics Education 19, no. 3 (2011). http://www.amstat.org/publications/jse/v19n3/decock.pdf.

Elwert, Felix. "Graphical Causal Models." In Handbook of Causal Analysis for Social Research, edited by Stephen L. Morgan. Handbooks of Sociology and Social Research. Dordrecht: Springer Netherlands, 2013. http://link.springer.com/10.1007/978-94-007-6094-3.

Elwert, Felix, and Christopher Winship. "Endogenous Selection Bias: The Problem of Conditioning on a Collider Variable." Annual Review of Sociology 40, no. 1 (July 30, 2014): 31–53. doi:10.1146/annurev-soc-071913-043455.

Enders, Craig K. Applied Missing Data Analysis. 1 edition. New York: The Guilford Press, 2010.

Holland, Paul W. "Statistics and Causal Inference." Journal of the American Statistical Association 81, no. 396 (1986): 945–60.

"International Association for Impact Assessment." Accessed October 19, 2014. http://www.iaia.org/about/.

Koller, Daphne, and Nir Friedman. Probabilistic Graphical Models: Principles and Techniques. 1 edition. Cambridge, MA: The MIT Press, 2009.

Lauritzen, S. L., and D. J. Spiegelhalter. "Local Computations with Probabilities on Graphical Structures and Their Application to Expert Systems." Journal of the Royal Statistical Society. Series B (Methodological) 50, no. 2 (January 1, 1988): 157–224.

"Macmillan Dictionary | Free English Dictionary and Thesaurus Online." Accessed August 27, 2015. http://www.macmillandictionary.com/us.

Mangasarian, Olvi L, W. Nick Street, and William H Wolberg. "Breast Cancer Diagnosis and Prognosis via Linear Programming." OPERATIONS RESEARCH 43 (1995): 570–77.

Manski, Charles F. Identification Problems in the Social Sciences. Harvard University Press, 1999.

Pearl, Judea. Causality: Models, Reasoning and Inference. 2nd ed. Cambridge University Press, 2009.

Pearl, Judea, and Stuart Russell. "Bayesian Networks." UCLA Congnitive Systems Laboratory, November 2000. http://bayes.cs.ucla.edu/csl_papers.html.

Peugh, James L., and Craig K. Enders. "Missing Data in Educational Research: A Review of Reporting Practices and Suggestions for Improvement." Review of Educational Research 74, no. 4 (December 1, 2004): 525–56. doi:10.3102/00346543074004525.

Rubin, Donald B. "Estimating Causal Effects of Treatments in Randomized and Nonrandomized Studies." Journal of Educational Psychology 66, no. 5 (1974): 688–701. doi:10.1037/h0037350.

Russell, Stuart J. "Judea Pearl - A.M. Turing Award Winner." Accessed July 6, 2014. http://amturing.acm.org/award_winners/pearl_2658896.cfm.

Shmueli, Galit. "To Explain or to Predict?" Statistical Science 25, no. 3 (August 2010): 289–310. doi:10.1214/10-STS330.

Shpitser, Ilya, Tyler Vanderweele, and James M. Robins. "On the Validity of Covariate Adjustment for Estimating Causal Effects." In In Proceedings of the 26th Conference on Uncertainty and Artificial Intelligence, Eds. P. Grunwald & P. Spirtes, 527–36, 2010.

Snee, Ronald D. "Graphical Display of Two-Way Contingency Tables." The American Statistician 28, no. 1 (February 1, 1974): 9–12. doi:10.2307/2683520.

Tuna, Cari. "When Combined Data Reveal the Flaw of Averages." Wall Street Journal, December 2, 2009, sec. US. http://www.wsj.com/articles/SB125970744553071829.

van Buuren, Stef. "Multiple Imputation of Discrete and Continuous Data by Fully Conditional Specification." Statistical Methods in Medical Research 16, no. 3 (June 2007): 219–42. doi:10.1177/0962280206074463.

Weber, Philippe, and Lionel Jouffe. "Reliability Modelling with Dynamic Bayesian Networks," 57–62. IFAC, 2003. https://hal.archives-ouvertes.fr/hal-00128475.

Wolberg, W. H., and O. L. Mangasarian. "Multisurface Method of Pattern Separation for Medical Diagnosis Applied to Breast Cytology." Proceedings of the National Academy of Sciences 87, no. 23 (December 1, 1990): 9193–96. doi:10.1073/pnas.87.23.9193.

Wolberg, W. H, W. N Street, D. M Heisey, and O. L Mangasarian. "Computer-Derived Nuclear Features Distinguish Malignant from Benign Breast Cytology* 1." Human Pathology 26, no. 7 (1995): 792–96.

Wolberg, William H, W. Nick Street, and Olvi L Mangasarian. "Breast Cytology Diagnosis Via Digital Image Analysis," 1993. http://citeseerx.ist.psu.edu/viewdoc/summary?doi=10.1.1.38.9894.

Wolberg, William, W. Nick Street, and Olvi Mangasarian. "Breast Cancer Diagnosis and Prognosis via Linear Programming," December 19, 1994. http://digital.library.wisc.edu/1793/64370.

Ευσεβιου του παμφιλου. Εὐαγγελικὴ προπαρασκευή (Praeparatio Evangelica). Patrologiae cursus completus. Paris, 1857.

Index

A

Acceptance Threshold 127
ACE. *See* Average Causal Effect (ACE)
Adaptive Questionnaire 45, 147, 148, 149, 151, 153, 154, 155, 156, 159
Adjustment Criterion 340, 341, 352
Approximate Inference 27
Arc Analysis 161
Arc Comments 104, 126, 246
Arc Creation Mode 54
Arc Force 181, 182, 183, 184, 185, 216
Arcs' Mutual Information 104, 108, 110, 125
Artificial Intelligence 15, 17
Associate Data Wizard 345, 346, 348
Associate Dictionary 92, 173, 264
Associate Evidence Scenario 278
Augmented Markov Blanket 39, 133, 134, 135, 136, 137, 139, 142, 143, 144
Automatic Layout 124, 178, 235, 245, 251
Average Causal Effect (ACE) 329, 331

B

Batch Inference 46
Bayes Factor 22, 196, 197
Bayesia Engine API 46
Bayesia Expert Knowledge Elicitation Environment (BEKEE) 36
Bayesian Updating 29, 38
Bayes' Theorem 21

Bias 88, 289, 292, 330, 332, 342, 343

Binary Algorithm 192, 193

Binary Mutual Information 148

Breakout Variable 267

C

Casewise Deletion 298, 304, 307

Causal Bayesian Network (CBN) 18, 23, 52, 290, 291, 292, 334, 344, 345

Causal DAG 335, 336, 338, 339, 342, 344

Causal Discovery 29

Causal Identification 325, 328, 330

Causal Inference 44, 325, 327

Causal Network 28, 30

Causal Path 339, 340

Causal Reasoning 27

CBN. *See* Causal Bayesian Network (CBN)

CDAG. *See* Causal DAG

Chain Rule. *See* Product Rule

Chi-Square Test 258

Class Editor 222, 247

Classes 115, 117, 122, 123, 124, 125, 126, 127, 146, 147, 148, 149, 150, 151, 152, 156, 158, 159, 160, 162, 221, 222, 247, 248, 253, 259

Classification 38, 45, 113, 114, 118, 144, 162

Cluster Cross-Validation 223, 227

Clustering 40, 203, 216, 217, 221, 222, 224, 225, 226, 227, 228, 229, 235, 239, 242, 243, 247, 248

Clustering Frequency Graph 226

Clustering Report 221, 224, 225, 239

Cluster Validation 221

Collider 337, 338

Common Cause 337, 339

Common Effect 29, 63, 337, 338, 339

Comparison Network 132, 135

Complexity 29, 39, 88, 138, 139, 141, 143, 176, 212, 313, 328, 336

Comprehensive Report 131

Conditional Entropy 98, 319

Conditional Mutual Information 160

Conditional Probability Distribution (CPD) 25, 26, 292, 294, 296

Conditional Probability Table (CPT) 22, 23, 26, 29, 36, 37, 44, 54, 58, 60, 61, 72, 99, 101, 102, 103, 120, 349

Conflicting Evidence 194

Console 213, 214, 216

Constraint-Based Algorithm 29

Contextual Menu 63, 66, 106, 142, 144, 185, 190, 222, 269, 314, 353

Contingency Table Fit (CTF) 240, 241, 243

Contributions 243, 268

Correlation 15, 18, 111, 112, 144, 145, 177

Correlation Coefficient 95

Cost 148, 149, 160, 262, 263, 264, 274, 277

Cost Editor 148, 262, 263, 264

Counterfactual 327, 329

Covariance 79, 95

CPD. *See* Conditional Probability Distribution (CPD)

CPT. *See* Conditional Probability Table (CPT)

Criterion Optimization 276

Cross-Entropy 191

Cross-Validation 129, 132, 136, 143

CTF. *See* Contingency Table Fit (CTF)

D

DAG. *See* Directed Acyclic Graph (DAG)

Data Clustering 40, 203, 228, 229, 235, 239, 242, 247

Data Clustering Report 239

Data-Generating Process (DGP) 290, 291, 292, 294, 295, 301, 312, 333, 334

Data Import 80, 81, 82, 84, 301, 302

Data Import Wizard xi, 36, 80, 81, 82, 84, 115, 116, 117, 168, 169, 170, 203, 290, 301, 302, 345, 346

Data Mining 16

Data Perturbation 215, 216, 224, 250

DBN. *See* Dynamic Bayesian Network

Decay Factor 215

Decision Node 354

Define Typing 116

Degree of Freedom 258

Dendrogram 219, 221

DGP. *See* Data-Generating Process (DGP)

Diagnosis 41, 114, 162

Dictionary 92, 93

Directed Acyclic Graph (DAG) 21, 22, 325, 334, 335, 336, 338, 339, 341, 343, 344, 351, 356

Directed Arc 336

Direct Effects 44, 360, 361, 362

Direct Effects Analysis 360, 361

Direct Effects on Target 360

Discretization 36, 84, 85, 86, 87, 88, 89, 90, 95, 97, 170, 205

Discretization Intervals 119, 120, 121

Display Arc Comments 104, 126

Display Horizontal Scales 268

Display Node Comments 93, 174

Display Sensitivity Chart 255, 363

Display Vertical Scales 268

Do-Operator 44, 354

Dynamic Bayesian Network 24, 25

Dynamic Imputation 37, 311, 312, 314, 315, 316, 317, 318, 319, 320, 321

E

Edit Costs 148, 263

Edit Structural Coefficient 142

Enter Probabilities 188, 358

Entropy 79, 88, 95, 96, 97, 98, 104, 105, 111, 112, 318, 319, 320

Entropy-Based Dynamic Imputation 318

Entropy-Based Static Imputation 318

EQ Algorithm 210, 214, 224, 249, 321

Equal Distance 87, 88, 206

Equal Frequency 88

Evidence Analysis Report 196, 197

Evidence Scenarios 278, 283

Evidential Reasoning 27, 60, 73

Explanatory Modeling 16

Export Variations 272

F

Factor 221, 222, 227, 228, 229, 233, 235, 236, 237, 238, 241, 242, 243, 245, 246, 247, 249, 252, 253, 254, 255, 258, 259, 260, 265

Factor Analysis 40, 227

Filter 83, 298, 301, 304, 305, 306, 307, 308, 309, 313

Filtered States 84

Filtered Values 81, 84, 90, 290, 296, 297, 298, 299, 300, 302, 303, 304

Fixed Arcs 249

Fix Mean 191

Fix Probabilities 191

Forbidden Arc Editor 248

Forbidden Arcs 248

G

Generate Data 299

Generate States 56

Global Conflict 196

Graph Panel 68, 69, 90, 142, 216, 222

Graph Surgery 352, 353

G-test 258

H

Hard Evidence 42, 159, 187, 188, 193, 278, 337

Hidden Markov Model (HMM) 25

Hide Information 109

HMM. *See* Hidden Markov Model (HMM)

Horizontal Distribution 314
Horizontal Scales 268, 269, 270
Hyperparameters 29, 38

I

Identification 325, 328, 330, 334, 339, 340
Ignorability 330, 331
Import Report 89, 90, 120, 121, 206
Indirect Effect 339
Individual-Level Causal Effect (ICE) 329
Influence Analysis on Target 66, 75
Influence Paths to Target 350
Information Panel 117, 302, 306, 308
Interactive Inference 145, 146, 238
Inter-Causal Reasoning 63
Intervention 353, 354, 355, 364
Intervention Mode 353, 355

J

JAR. *See* Just-About-Right (JAR) Variable
JavaScript 46
JDBC 80
Joint Probability 23, 25, 27, 28, 51, 120, 123, 157, 162, 181, 195, 276, 281
Joint Probability Distribution (JPD) 23, 25, 26, 27, 28, 51, 123, 181, 212, 216, 223, 228, 229, 240, 241, 299, 333, 358
JPD. *See* Joint Probability Distribution (JPD)
Just-About-Right (JAR) Variable 202, 260, 286

K

Kalman Filter 25
Key Drivers Analysis 253
K-Folds Cross Validation 129, 132, 136, 143

K-L Divergence. *See* Kullback-Leibler Divergence

K-Means Discretization 88

Kullback-Leibler Divergence 181

L

Learning Set 116, 130, 133

Likelihood 22

Likelihood Evidence 42

Likelihood Matching 44, 358, 360

Likelihood Ratio 22

Linearize Nodes' Values 267

Listwise Deletion 83, 298, 304, 307. *See also* Filter

Local Conflict 196

Local Consistency 196, 197

Log-Likelihood 212, 240, 241

M

Mapping 160, 161, 185, 219, 221, 233

MAR. *See* Missing at Random (MAR)

Marginal Entropy 98, 105

Markov Blanket 39, 123, 124, 125, 127, 130, 133, 134, 135, 136, 137, 139, 142, 143, 144, 319

Matching 355, 356

Maximum Likelihood 29, 37, 38, 91, 99, 237

Maximum Likelihood Estimation 37, 38

Maximum Size of Evidence 157

Maximum Spanning Tree. *See* Maximum Weight Spanning Tree (MWST)

Maximum Weight Spanning Tree (MWST) 176, 177, 210, 320

MCAR. *See* Missing Completely at Random (MCAR)

MDL. *See* Minimum Description Length

MDL Score 211, 213, 214, 241, 251. *See also* Minimum Description Length

MI. *See* Mutual Information

Minimum Cross-Entropy 191

Minimum Description Length 38, 177, 212

Minimum Joint Probability 157, 281

MinXEnt 191, 193, 194, 195

Missing Arc 336

Missing at Random (MAR) 290, 293, 294, 297, 311, 312

Missing Completely at Random (MCAR) 290, 291, 292, 305, 306, 307

Missing Not at Random (MNAR) 290, 295, 296, 311, 312, 315

Missing Values 37, 81, 83, 90, 117, 170, 289, 298, 301, 303, 305, 321

Modeling Mode 67, 134, 142, 170, 172, 176

Monitor Panel 61, 64, 144, 145, 300

Monte Carlo Simulation 27

Multiple Clustering 40, 203, 227, 228, 242, 243, 248

Multiple Discretization 118

Multi-Quadrant Analysis 266, 269, 273, 278

Mutual Information 95, 98, 103, 104, 105, 106, 107, 108, 109, 110, 111, 112, 125, 126, 145, 148, 149, 160, 161, 181, 233, 234, 235

Mutual Information with the Target Node 126, 161

MWST. *See* Maximum Weight Spanning Tree (MWST)

N

Naive Bayes Network 39, 105, 106, 107

Navigation Bar 146

Network Analysis 179

Network Comparison 132

Network Performance 127, 239

Node Analysis 161

Node Comments 92, 93, 172, 173, 174, 246, 251, 257

Node Editor 36, 53, 54, 56, 57, 58, 60, 72, 92, 102, 103, 122, 264, 349

Node Exclusion 207

Node Force 183, 184, 185, 220

Node Force Mapping 185

Node Names 69, 92, 172, 173, 257

Node Property 262

Non-Confounder 361

Non-Descendant 26

Normalize 57
Normalized Equal Distance 88
Not Missing at Random. *See* Missing Not at Random (MNAR)
Not Observable 262, 263, 265, 277
Numerical Evidence 42, 188, 190, 192, 193, 194

O

Observational Inference 37, 41
Occurrences 91
OLS. *See* Ordinary Least Squares (OLS)
Omnidirectional Inference 62
Ordinary Least Squares (OLS) 328, 342
Overall Conflict 196, 197
Overfitting 139

P

Parameter Estimation 37, 99, 102, 103, 267, 320, 321, 345, 348
Parents 22, 25, 26, 28, 119, 120, 124, 181, 292, 294, 296, 298, 352
Path Analysis 349
PCA. *See* Principal Components Analysis (PCA)
PDAG. *See* Probabilistic DAG
Pearson's Correlation 112, 177
PHP 46
Potential Outcomes Framework 329
Predictive Modeling 16
Principal Components Analysis (PCA) 227
Probabilistic DAG 335, 336
Probabilistic Evidence 42, 188, 190, 358
Probabilistic Structural Equation Model (PSEM) xi, 40, 201, 202, 241, 248, 253, 267
Probability Distribution 53, 57, 91, 103
Product Rule 26, 27, 240
Progression Margin 279, 280
PSEM. *See* Probabilistic Structural Equation Model (PSEM)

Q

Quadrant Plot 235, 268

Quadrants 234

Questionnaire Target 151, 152

R

Radial Layout 109, 110, 126

Rediscretize Continuous Nodes 267

Reference Network 132, 133, 135

Regenerate Values 267

Relationship Analysis 111, 112

Relative Mutual Information 104, 105

Relative Significance 233, 236

Replace By 301, 308, 309, 310, 313

Row Identifier 117

S

SAS 46

SC. *See* Structural Coefficient (SC)

Score 160

Search Stop Criteria 157

Set as Target Node 65, 106, 237

Shuffle Samples 130

Simpson's Paradox xi, 332, 333, 338, 343, 364

Soft Evidence 42, 193, 275, 278, 280. *See also* Probabilistic Evidence

SopLEQ 210

Standardized Total Effect 258

State Names 122

Static Imputation 310, 312, 313, 318, 319, 320, 321

Stretch 109

Structural Coefficient Analysis 139, 142

Structural Coefficient (SC) 138, 139, 140, 141, 142, 212

Structural EM 37, 83, 117, 316, 317, 321. *See* Structural Expectation Maximization

Structural Equation Model 201, 248, 253, 289

Structural Expectation Maximization. *See* Structural EM

Structure Comparison 135

Structure/Target Precision Ratio 141

Supervised Learning xi, 39, 88, 89, 107, 113, 123, 133, 134, 181

Switch Position 158

Synthesis Structure 132

T

Taboo 210, 249, 313

Taboo Order 210

Target Analysis 254, 256, 275, 360

Target Analysis Report 256

Target Correlation 144

Target Distribution 191, 193

Target Dynamic Profile 44, 274, 275

Targeted Evaluation 129, 136

Target Interpretation Tree 156, 157, 158, 162

Target Mean Analysis 44, 66, 254, 259, 362

Target Mean/Value 190, 191, 192, 194

Target Node 65, 66, 105, 106, 107, 109, 123, 125, 126, 127, 144, 145, 146, 148, 149, 150, 151, 160, 161, 181, 203, 207, 233, 238, 248, 249, 254, 255, 257, 258, 268, 276, 280, 281, 349, 361

Target Optimization 44

Target State 106, 148, 160

Test Set 116, 127, 128, 129, 130, 132, 138, 143

Time Series 168

Total Effects 44, 256, 257, 258, 259, 260, 267, 268, 271, 355

Total Effects on Target 256, 259, 260, 354, 355

Total Precision 127, 128, 131

U

Uncertainty 41, 42, 49, 50, 51, 95, 96, 97, 104, 105, 166, 187, 188, 192, 193, 194, 195,

202, 206, 319, 337, 338

Unobserved Variables 341

Unsupervised Learning xi, 38, 88, 89, 113, 133, 165, 170, 176, 181, 197, 198, 203, 207, 208, 211, 249, 313

Unsupervised Structural Learning. *See* Unsupervised Learning

Use of Classes 253, 259

Utilize Evidence Cost 160, 277

V

Validate Clustering 221, 227

Validation Mode 60, 73, 94, 103, 104, 108, 125, 144, 175, 181, 183, 185, 216, 236, 253, 254, 275

Value Shift 192, 193

Variable Clustering 40, 203, 216, 217, 221, 222, 224, 225

Variable Clustering Report 224, 225

Variation Editor 277, 278, 279

VBA 46

Vertical Scales 268, 269, 270

Virtual Evidence 42

V-Structure 337

W

WebSimulator 45, 46, 151, 152, 153

WebSimulator Editor 151

Made in the USA
Lexington, KY
10 October 2016